*Thomas Heine, Jan-Ole Joswig,
and Achim Gelessus*

Computational Chemistry Workbook

Related Titles

J.R. Fried

Computational Chemistry and Molecular Simulation

2010
ISBN: 978-0-471-46244-6

P. Comba, T.W. Hambley, B. Martin

Molecular Modeling of Inorganic Compounds

2009
ISBN: 978-3-527-31799-8

T. Kinzel, F. Major, C. Raith, T. Redert, F. Stecker,
N. Tölle, J. Zinngrebe

Organic Synthesis Workbook III

2007
ISBN: 978-3-527-31665-6

B.M. Rode, T. Hofer, M. Kugler

The Basics of Theoretical and Computational Chemistry

2007
ISBN: 978-3-527-31773-8

Thomas Heine, Jan-Ole Joswig,
and Achim Gelessus

Computational Chemistry Workbook

Learning Through Examples

With a Foreword by
Dennis R. Salahub

WILEY-VCH

WILEY-VCH Verlag GmbH & Co. KGaA

The Authors

Prof. Dr. Thomas Heine
Jacobs University Bremen
School of Engineering and Science
Campus Ring 1
28759 Bremen
Germany

Dr. Jan-Ole Joswig
Technical University Dresden
Physical Chemistry
Bergstrasse 66 b
10169 Dresden
Germany

Dr. Achim Gelessus
Jacobs University Bremen
School of Engineering and Science
Campus Ring 1
28759 Bremen
Germany

1st Reprint 2011

Library of Congress Card No.: applied for

British Library Cataloguing-in-Publication Data
A catalogue record for this book is available from the British
Library.

**Bibliographic information published by
the Deutsche Nationalbibliothek**
The Deutsche Nationalbibliothek lists this
publication in the Deutsche Nationalbibliografie; detailed
bibliographic data are available on the
Internet at http://dnb.d-nb.de.

© 2009 WILEY-VCH Verlag GmbH & Co. KGaA, Weinheim

Printed in the Federal Republic of Germany
Printed on acid-free paper

Cover Design   Formgeber, Eppelheim
Typesetting   Thomson Digital, Noida, India
Printing   Strauss GmbH, Mörlenbach
Bookbinding   Litges & Dopf Buchbinderei GmbH,
Heppenheim

ISBN: 978-3-527-32442-2

Dedicated to Prof. Dr. Gotthard Seifert

Foreword

This *Computational Chemistry Workbook* nicely bridges a gap in the teaching and learning of quantum chemical concepts and their applications. Many textbooks treat the usual array of "solvable" problems, the hydrogen atom, the harmonic oscillator, simple Hückel examples, etc., with varying degrees of mathematical rigor. But the list of such solvable problems is relatively short, so that one often gets the impression that there are two worlds, the small, solvable problems of the textbooks, on the one hand, and the real world of chemical experiments, on the other.

This *Workbook* brings in the computational component, to complement formal theory and build a better bridge to experiment. With the accompanying CD, anyone who wants to design a computational chemistry laboratory course now has a very clearly written manual, plus all of the necessary software. All that is needed is to boot the CD on any type of PC and a wide variety of interesting problems can be solved. The student can adapt the exercises very widely, to explore personal interests within whatever kind of chemistry is of interest to him or her. The onus is on using the methods on the CD, rather than the details of how to solve Schrödinger's equation.

Examples treated include: molecular coordinates and symmetry (mainly to prepare input to an electronic structure code), vibrational states and the spectroscopy of diatomic molecules, atomic orbitals, ionization potentials, electron affinities, Hückel molecular orbital theory, atomic charges, bond orders, geometry optimization of a diatomic molecule, electron spin, vibrational spectroscopy, and, for more advanced students, thermochemistry and the important link to statistical mechanics and molecular dynamics, including "on-the-fly" or Born–Oppenheimer molecular dynamics. The range is broad, but, at each step, the authors have taken pains to ensure a careful explanation of the basics, and enough details so that the student can then improvise and explore.

Computational Chemistry Workbook: Learning Through Examples
Thomas Heine, Jan-Ole Joswig, and Achim Gelessus
Copyright © 2009 WILEY-VCH Verlag GmbH & Co. KGaA, Weinheim
ISBN: 978-3-527-32442-2

I am very happy that the authors have chosen to use our deMon density functional software to drive the exercises, though the framework is flexible enough to allow other programs to be used as well. deMon has been developed over the years by a loose international consortium of enthusiastic and dedicated researchers known as the deMon Developers (see http://demon-software.com). I am sure that deMon will serve the needs of this *Workbook* well, although, of course, we would be happy to hear of any difficulties that might be encountered.

Congratulations on this very significant contribution to quantum chemical education!

Dennis R. Salahub
PhD, FRSC, FAAAS
University of Calgary

Acknowledgments

Many people have contributed to this workbook, to all of whom the authors are deeply indebted. We have been designing these computer experiments for several years for chemistry students in their second, third and fourth years at Dresden University of Technology. In particular, when the curriculum was reformed and – as a result of the Bologna Process – Bachelor and Master programs in chemistry were established, we took the opportunity to restructure computational chemistry education.

First of all, we would like to thank Prof. Dr. Gotthard Seifert (TU Dresden) for his help and support in designing the computer experiments in close collaboration with his lectures in theoretical and computational chemistry, and for his encouragement to publish the resulting computer experiments. We are also deeply indebted to DP Knut Vietze (TU Dresden) for his invaluable technical support, the scientific discussions, and his didactic input, in particular in the early stages of this project. We are grateful to Dr. Robert Barthel (Marseille) for his help in the composition of the thermochemistry experiment.

A number of undergraduate students have contributed to this book by careful reading of the manuscripts and by testing the experiments on their home computers. We greatly appreciate their help, and we would especially like to thank Cvetomir Dimov, Joseph Kipriotich, and the Computational Materials Science class of 2008, as well as Florian Neu and the Computational Laboratory for Analysis, Modeling and Visualization (CLAMV) from Jacobs University and Jan Brückner (third year) from Dresden University of Technology. Special thanks go to Tommy Lorenz (diploma student, TU Dresden), who was strongly involved in planning some of the experiments when he was still an undergraduate student.

Finally, computational chemistry is not possible without computer codes. All the software on the accompanying CD is non-commercial and free for academic use. We are very grateful to all the often

Computational Chemistry Workbook: Learning Through Examples
Thomas Heine, Jan-Ole Joswig, and Achim Gelessus
Copyright © 2009 WILEY-VCH Verlag GmbH & Co. KGaA, Weinheim
ISBN: 978-3-527-32442-2

unnamed people who have made major or minor contributions to the software just for the benefit of members of the community. To name only a few, we are indebted to the developers of `Knoppix`, the `GFortran` compiler, `Linux`, `deMon`, `CaGe`, `molden`, `Xmgrace`, and `GNUplot`.

Thomas Heine
Jan-Ole Joswig
Achim Gelessus

Contents

Foreword   *VII*
Acknowledgments   *IX*

1	**Introduction** *1*	
1.1	Basics of Computational Chemistry *1*	
1.2	Aim of This Book *2*	
1.3	How to Use This Book *3*	
1.4	Structure of This Book *4*	
1.5	The Chapters *5*	
1.6	The Software *9*	
2	**Molecular Coordinates and Symmetry** *11*	
2.1	Aim *11*	
2.2	Theoretical Background *11*	
2.2.1	Cartesian and Internal Coordinates *11*	
2.2.2	The *Z*-Matrix *12*	
2.2.3	The Dihedral Angle *13*	
2.2.4	Symmetry *15*	
2.2.5	Symmetry Elements and Operations *16*	
2.2.6	Point Groups *17*	
2.3	Demonstration *19*	
2.3.1	Constructing a *Z*-Matrix *19*	
2.3.2	Determining Molecular Point Groups *20*	
2.4	Problems *22*	
2.5	Technical Details *23*	
2.6	Review and Summary *23*	
	References *23*	

Computational Chemistry Workbook: Learning Through Examples
Thomas Heine, Jan-Ole Joswig, and Achim Gelessus
Copyright © 2009 WILEY-VCH Verlag GmbH & Co. KGaA, Weinheim
ISBN: 978-3-527-32442-2

3   **Vibrations of Diatomic Molecules: The Harmonic Approximation** *25*
3.1   Aim   *25*
3.2   Theoretical Background   *25*
3.2.1   The Harmonic Approximation   *25*
3.2.2   The Reduced Mass   *27*
3.2.3   The Vibrational Frequencies   *28*
3.3   Demonstration   *30*
3.4   Problems   *32*
3.5   Technical Details   *33*
3.6   Review and Summary   *34*
        References   *34*

4   **Vibrations of Diatomic Molecules: The Schrödinger Equation** *35*
4.1   Aim   *35*
4.2   Theoretical Background   *35*
4.2.1   Classical Versus Quantum Mechanics   *35*
4.2.2   The Stationary Schrödinger Equation   *37*
4.2.3   Solutions to the Schrödinger Equation   *39*
4.3   Demonstration   *43*
4.4   Problems   *44*
4.5   Technical Details   *45*
4.6   Review and Summary   *45*
        References   *45*

5   **Atomic Orbitals**   *47*
5.1   Aim   *47*
5.2   Theoretical Background   *47*
5.2.1   The Schrödinger Equation of the Atom   *47*
5.2.2   Atomic Orbitals   *49*
5.3   Demonstration   *51*
5.4   Problems   *53*
5.5   Technical Details   *54*
5.6   Review and Summary   *54*
        References   *55*

6   **Ionization Potentials and Electron Affinities of Atoms**   *57*
6.1   Aim   *57*
6.2   Theoretical Background   *57*
6.2.1   Ionization Potential and Electron Affinity   *57*
6.2.2   Slater Rules: Wavefunction and Binding Energies of Electrons in Atoms and Ions   *59*
6.2.3   Calculation of Ionization Potentials and Electron Affinities   *61*

6.3   Demonstration   *62*
6.4   Problems   *64*
6.5   Technical Details   *65*
6.6   Review and Summary   *66*
        References   *66*

7   Hückel Molecular Orbital Theory:
        Stability of Conjugated Carbon π Systems   *67*
7.1   Aim   *67*
7.2   Theoretical Background   *67*
7.2.1   Molecular Orbital Theory   *68*
7.2.2   The Hückel Postulates   *70*
7.2.3   Topology Matrices   *71*
7.2.4   Values for α and β *73*
7.2.5   The Trap of Defining x *73*
7.2.6   π Electron Binding Energy   *74*
7.2.7   π Electron Molecular Orbitals and Probability Density   *74*
7.3   Demonstration   *75*
7.3.1   Hückel Calculation of the Cyclopropenyl Cation   *75*
7.4   Problems   *78*
7.5   Technical Details   *79*
7.6   Review and Summary   *80*
        References   *81*

8   Hückel Molecular Orbital Theory:
        Bond Order, Charge Order, and Molecular Orbitals   *83*
8.1   Aim   *83*
8.2   Theoretical Background   *83*
8.2.1   Bond Order   *83*
8.2.2   Charge Order   *84*
8.3   Demonstration   *85*
8.3.1   Hückel Calculation of the Butadiene Molecule   *85*
8.4   Problems   *88*
8.5   Review and Summary   *89*
        References   *90*

9   Geometry Optimization of a Diatomic Molecule   *91*
9.1   Aim   *91*
9.2   Theoretical Background   *91*
9.2.1   The Potential Energy Surface   *91*
9.2.2   Forces in a Diatomic Molecule   *93*
9.2.3   The Steepest Descent Method   *94*
9.2.4   Hessian-Based Optimizers and the Newton–Raphson
        Method   *95*
9.2.5   Application to the Morse Potential   *97*

9.2.6   Global Versus Local Geometry Optimization   *97*
9.3   Demonstration   *98*
9.3.1   Optimizing CO With the Steepest Descent
          Method   *98*
9.3.2   Optimizing CO Using a Hessian-Based Newton–Raphson
          Optimizer   *100*
9.4   Problems   *103*
9.5   Technical Details   *103*
9.6   Review and Summary   *104*
          References   *105*

10   The Electron Spin   107
10.1   Aim   *107*
10.2   Theoretical Background   *107*
10.2.1   The Electron Spin   *107*
10.2.2   The Multiplicity   *108*
10.2.3   The Jahn–Teller Effect   *108*
10.3   Demonstration   *109*
10.4   Problems   *110*
10.5   Technical Details   *111*
10.6   Review and Summary   *112*
          References   *112*

11   Vibrational Spectroscopy   113
11.1   Aim   *113*
11.2   Theoretical Background   *113*
11.2.1   Analysis of Classical Vibrations Within the Harmonic
          Approximation   *114*
11.2.2   The Harmonic Oscillator Revisited   *115*
11.2.3   The Vibrational Modes   *117*
11.2.4   Intensities   *119*
11.3   Demonstration   *120*
11.3.1   The Vibrational Modes of a Linear Molecule   *120*
11.4   Problems   *124*
11.5   Technical Details   *125*
11.6   Review and Summary   *126*
          References   *126*

**12   Vibrational Spectroscopy and Character Tables –
          Advanced Topics   127**
12.1   Aim   *127*
12.2   Theoretical Background   *127*
12.2.1   The Hessian Matrix   *127*
12.2.2   Normal Modes   *129*
12.2.3   Symmetry in Normal Modes   *130*

12.2.4   Selection Rules   *133*
12.3   Demonstration   *136*
12.4   Problems   *138*
12.5   Review and Summary   *140*
          References   *141*

**13   Ionization Potential and Electron Affinities
          of Molecules**   *143*
13.1   Aim   *143*
13.2   Theoretical Background   *143*
13.2.1   Field of Application   *143*
13.2.2   Influence of Geometry   *144*
13.3   Demonstration   *146*
13.4   Problems   *147*
13.4.1   Technical Details   *148*
13.5   Review and Summary   *148*
          References   *149*

14   Thermochemistry   *151*
14.1   Aim   *151*
14.2   Theoretical Background   *151*
14.2.1   Calculating Thermodynamic Functions Using the Partition
          Function   *152*
14.2.2   Thermochemistry Within the Ideal-Gas
          Approximation   *154*
14.2.3   The Molecular Partition Function q *156*
14.2.4   Calculating the Relative Abundance of Isomers in the
          Gas Phase   *160*
14.3   Demonstration   *161*
14.4   Problems   *164*
14.5   Review and Summary   *165*
          References   *165*

15   Molecular Dynamics – Basic Concepts   *167*
15.1   Aim   *167*
15.2   Theoretical Background   *167*
15.2.1   Computer Simulations in Chemistry   *167*
15.2.2   The Born–Oppenheimer Approximation   *169*
15.2.3   The Trajectory: Startup Conditions and Propagation of
          Atoms in Time   *169*
15.2.4   The Verlet Algorithm   *170*
15.2.5   The Velocity Verlet Algorithm   *171*
15.2.6   Conservation of Fundamental Physical Quantities:
          Energy, Momentum, and Angular Momentum   *172*
15.2.7   Numerical Considerations   *173*

15.2.8   Heat Bath: Thermostats   *174*
15.3   Demonstration   *174*
15.3.1   Classical Molecular Dynamics in a Spreadsheet   *174*
15.4   Problems   *178*
15.5   Review and Summary   *178*
            References   *179*

16   Molecular Dynamics and Basic Thermodynamics   *181*
16.1   Aim   *181*
16.2   Theoretical Background   *181*
16.2.1   Basic Thermodynamics: Ensembles   *182*
16.2.2   The Temperature and the Ergodic Theorem   *183*
16.2.3   The Connection to Real Macroscopic Systems   *185*
16.2.4   External Heat Baths – Thermostats   *186*
16.2.4.1   A Simple Scaling Thermostat   *186*
16.2.4.2   The Berendsen Thermostat   *186*
16.2.4.3   The Andersen Thermostat   *187*
16.2.4.4   The Nosé-Hoover Thermostat   *187*
16.2.5   Averages   *188*
16.3   Demonstration   *188*
16.4   Problems   *194*
16.5   Review and Summary   *195*
            References   *196*

17   Molecular Dynamics – Simulated Annealing   *197*
17.1   Aim   *197*
17.2   Theoretical Background   *197*
17.2.1   The Potential Energy Surface   *199*
17.2.2   Simulated Annealing   *200*
17.3   Demonstration   *201*
17.3.1   Inspecting Stationary Points With Simulated
            Annealing   *201*
17.3.2   Finding the Global Minimum of Al_4^{2-} Using Simulated
            Annealing   *203*
17.4   Problems   *205*
17.5   Review and Summary   *206*
            References   *207*

Appendix
The Computational Chemistry Software Delivered
with This Book   *209*
A.1   Getting Started   *210*
A.1.1   Booting Directly the Live System   *210*
A.1.2   Booting as Virtual Machine   *210*
A.1.3   Running From a Pen Drive (USB Stick)   *211*

A.1.4   Configure Your Environment   *213*

A.2   A Brief Introduction to Linux   *214*

A.3   Character Tables for Chemically Important Point
         Groups   *218*

A.4   Computational Chemistry Software Delivered With
         This Book   *219*

A.4.1   Molden   *219*

A.4.2   deMon   *219*

A.4.3   CaGe   *220*

A.4.4   hueckel   *222*

A.4.5   THERMO Scripts   *222*

A.4.6   Xmgrace   *223*

A.4.7   GNUplot   *223*

Index   *225*

1
Introduction

1.1
Basics of Computational Chemistry

Computational chemistry has become an indispensable tool for the experimental chemist. Today, computational chemistry is used by the chemistry community to back up experimental results not only in many scientific publications, but also in textbooks and theses. The level of involvement varies strongly. Computational chemistry is, for example, employed to interpret and rationalize results, to compare experimental with simulated data, to get an idea of a molecular structure, or just to create a graphical representation of a molecule, its molecular orbitals or its electrostatic potential. Though only a small percentage of all chemists would call themselves computational, quantum or theoretical chemists, computational chemistry is indeed applied by the majority of modern chemists. This is due to the tremendous progress that the field has made in the past years. In contrast to the time a few decades ago, today we have powerful computers and – probably more importantly – modern, easy-to-use and robust chemistry software.

Progress in computational chemistry has allowed a new quality of research in chemistry. Most modern chemists actively use the computer at various levels of their work, for example to get an idea of the three-dimensional structure of molecules of interest, and to understand chemical bonding, stability and so on. It is now possible to calculate the stability of a molecule, or even a reaction path, before going into the laboratory to try the synthesis in practice. Also, it is common practice to simulate properties and to compare them with experiment, most importantly for infrared (IR), Raman and nuclear magnetic resonance (NMR) spectra.

Still, this exciting development comes with a dark side. Very often, the calculations are performed using the "black-box" strategy: well-established recipes are used to obtain a computational result of a particular problem, while very little is known about what exactly happens in the computer simulation. This unsatisfactory state of affairs

Computational Chemistry Workbook: Learning Through Examples
Thomas Heine, Jan-Ole Joswig, and Achim Gelessus
Copyright © 2009 WILEY-VCH Verlag GmbH & Co. KGaA, Weinheim
ISBN: 978-3-527-32442-2

is not the fault of today's experimental chemists. While the theoretical chemist receives all the education that is necessary to understand laboratory synthesis or analytics, sufficient theoretical chemistry education does not reach students who concentrate on experiments. We think that the reason for this problem stems from the history of computational chemistry: it is necessary to know a great deal of higher mathematics to develop quantum chemical methods, and a lot of programming, programming languages, and computer hardware to write efficient computer algorithms. Most chemists are not used to apply higher mathematics and are not trained in computer science or advanced quantum physics. This makes communication between experimentalists and theorists difficult.

Furthermore, when lecturing computational chemistry, it is difficult if not impossible to find typical "example problems" to practice what has been taught in class. Computer laboratory courses, where students have the chance to solve example problems on computers, are a way to bridge this gap. Again, courses for training the fundamental methods and algorithms that are used in a computer simulation are rare, and laboratory courses are usually unique, developed locally to accompany a certain lecture. In worst cases they are even restricted in providing technical instructions for the use of some special software.

1.2
Aim of This Book

The headmost aim of this workbook is to reduce this gap. We have demonstrated in several years at Dresden University of Technology that second- and third-year chemistry students can understand the basic principles of computational chemistry and create a solid basis to advanced quantum and computational chemistry education.

We do not expect that an undergraduate chemistry student will master tasks that are taught to physicists in their Master courses, such as many-body quantum mechanics, or which involve advanced mathematics, such as how to solve multi-dimensional partial differential equations. Instead, we avoid the whole topic of how the electronic Schrödinger equation of a molecule has to be solved. This is a subject for later studies. We concentrate instead on the concepts that are necessary to understand many phenomena in chemistry, and must be understood before computational chemistry can be used effectively in a more elaborate way. Whenever mathematics is necessary – and this is quite often the case – we try to be as explicit as possible and to illustrate the procedure to arrive at the result in great – for some people's taste probably too great – detail. However, we are sure that many students will appreciate these efforts, since they are often skipped or implied in many textbooks. Certain examples are carried out using spread-sheets. They allow one to apply geometry optimization or molecular dynamics algorithms, and students, even those without any programming knowledge, are able to "program" an algorithm for doing important tasks of computational chemistry.

This book comes with a CD, containing a full `Linux` operating system based on the `Knoppix` Live system, enriched with various computer programs that allow the computer experiments to be practiced on any reasonable Intel x86-compatible computer that can boot from a CD drive (Windows, Mac OS, `Linux`). It is also possible to work more effectively by running the CD as guest operating system in a virtual machine, by copying it to a pen drive, or by copying the book-related software to an existing `Linux` installation. This allows a straightforward application of the software in any computer pool or on any laptop or office PC without disturbing the local hard- and software setup. File transfer is easily possible through USB data storage media or the Internet. The system is kept up to date on a web site accompanying this project at http://www.compchem.jacobs-university.de/workbook.html.

It is, however, not necessary to use the software delivered with this book. All chapters are formulated in a neutral way and can be solved by employing different software to the taste, for the convenience and within the budget of the reader.

1.3
How to Use This Book

This book can be used in various ways. *Instructors* may take advantage of the book to design quickly a computational chemistry laboratory course. The book comes with all the software necessary to run the experiments, and the accompanying CD can be installed in any computer laboratory, even on the fly, leaving the standard system untouched. It might also be a convenient basis to set up some selected computer laboratory sessions, which may accompany courses in general chemistry, physical chemistry, but possibly also in analytic, organic or inorganic chemistry. The computer experiments can be easily adopted to include different example systems. It might also be interesting to distribute selected chapters – or modifications of it – as homework hands-on tasks: the individual chapters are self-contained, and detailed instruction is given to be able to solve the problems. To assist this purpose, the accompanied CD can be redistributed without charge, and newer versions will be available through the Internet.

For *students*, the book tries to close the gap between theory lectures and textbooks on the one hand and applications in typical "black-box computer programs" on the other, where it is hard to understand what exactly happens in the computer calculations. This book offers the possibility to practice various topics learned in class and to take the first steps in using professional computational chemistry software. The experiments may invite one to use computational chemistry in more advanced situations, and the distributed methods and software are a solid basis to do so.

1.4
Structure of This Book

Each chapter contains one type of computer experiment and is structured in sections following the same pattern:

- **Aim:** The aim of the computer experiment is briefly described. Here you can find the methods that will be introduced as well as the chemical problems that will be tackled.

- **Theoretical background:** As far as necessary the theory needed for the chapter will be reviewed. Bibliography to milestone papers in the original literature are given as well as the recommendations of standard textbooks. As this workbook is not intended to replace a standard text, we cover only the part of the theory that is relevant for the computer experiment, and sometimes we replace complex discussions that are not appropriate at this level by figurative explanations.

- **Demonstration:** A particular example is chosen and its solution is demonstrated in detail. If possible, the example is simplified to an extent that it can be treated by the student using only pencil and paper. Afterwards, the example is repeated using computer tools. After this demonstrative example, the reader is familiar with the practical realization of the theoretical concepts.

- **Problems:** This is the most important section. First, the hands-on tasks will be given. Then, general hints for their solution are provided. These hints are general recommendations to solve the hands-on tasks and do not cover any particular features of specific software. Technical assistance to solve the hands-on tasks using the specific software delivered with this book is then given in the "Technical Details". It is, however, straightforward to solve the problems using various different software products other than those found on the CD.

- **Review and summary:** Here, the computer experiment is summarized and its key elements are discussed briefly.

- **Bibliography:** This gives a general list of milestone papers in the original literature and the recommendations of standard textbooks and a selection of chapters therein.

1.5
The Chapters

The book contains 16 chapters of computational chemistry "experiments", each designed as a laboratory session to cover a certain topic. The Appendix gives technical advice on the use of the CD that is distributed with this book. Further technical information material can be found on the CD, which is regularly updated and available for download. The 16 chapters can be worked through successively. However, in many cases, specific chapters may be useful to accompany a certain course. Here, we give a recommendation on the context of the usage of each chapter.

Chapter 2 is an introductory chapter in which students learn how to communicate molecular structures with the computer. Essential software such as a molecular editor will be introduced. In this vein, the Cartesian and internal coordinates (Z-matrix) will be introduced. We take advantage of the visualization of molecular structures and introduce symmetry elements and point groups.

If the content of Chapter 2 is not yet known from an earlier introductory computational chemistry lab course, this chapter is highly recommended as a starting point. It also provides exercises for the introduction of molecular symmetry in a general chemistry or physical chemistry freshmen course.

Chapters 3 and 4 are one unit, introducing the computational approach to calculate the vibrational frequencies of diatomic molecules. In Chapter 3, we introduce the Morse potential to describe the potential energy surface (PES) of the diatomic molecule. We motivate the harmonic approximation to calculate the vibrational frequencies, and use the concept of reduced mass to solve the resulting differential equation. After obtaining the vibrational frequencies, we discuss the relative mobility of the two nuclei of the diatomic molecule depending on their relative masses. Chapter 3 can be skipped if all those concepts are already clear to the student. Chapter 4 starts by discussing the limitations of the classical approach of Chapter 3, and introduces the stationary Schrödinger equation to solve the motion of the nuclei on the PES of a diatomic. The important concepts of quantum mechanics are introduced here: quantization, the quantum mechanical state, the discrete set of eigenenergies and corresponding eigenfunctions, probability densities, and tunneling. Students understand that only discrete amplitudes are allowed for molecular vibrations, and that light quanta (photons) of the same frequency as the molecular vibrations are needed to excite a vibration or to emit a photon. Finally, the zero-point energy is introduced and discussed.

These two experiments are helpful for students who are confronted with quantum mechanics for the first time, for example in a physical chemistry course, and have a limited background in mathematics and in particular in quantum physics. In some cases Chapters 3 and 4 may be combined into a single practical course, or as a homework to refresh the knowledge of physics (Chapter 3), and to prepare for the lab course in Chapter 4.

In **Chapter 5** the electronic stationary Schrödinger equation is solved for some noble gas atoms. The typical textbook example, the analytical solution of the Schrödinger equation of hydrogen, is skipped, due to the involved mathematics. Instead, we state at this point that the exact solution of the electronic Schrödinger equation is impossible for any realistic system, that high-precision numerical approaches are very sophisticated, and that we will not look into this matter within this book. However, we are aware that there are means of computational chemistry that are able to approximate the solution of the electronic Schrödinger equation of a molecule, and that we will use such tools here and in the following. Details can be taught to an audience with stronger mathematical background, for example in a more advanced lecture or text. At this stage of education, however, it is important to develop a figurative understanding of the solution of the electronic Schrödinger equation. We discuss in detail the results of the calculation, in particular the orbital energies and the molecular orbitals, and allow the student to gain an insight into the energy and length scales for the core area and the valence area of atoms. We allow the student to visualize atomic orbitals using graphical software, to see nodes, orbital shapes, and the like. Finally, we request some mathematical skills by solving some calculus problems, where students show that atomic orbitals are normalized and orthogonal to each other.

Chapter 5 is designed to suit two purposes. It contains a lot of technical information, which will be necessary later. It also allows a better understanding of the spatial form of atomic orbitals, the extensions of atoms, nodes, and so on, to be obtained. We recommend it as a training tool for the quantum chemistry software. It might be combined with Chapter 1 for more advanced students to provide the technical background for further experiments.

In **Chapter 6**, the ionization potential and electron affinity of a selection of atoms will be determined. To understand the important concept of electron screening, we start the discussion with the Slater rules. We also use the figuratively clear scheme of Koopmans' approximation to the ionization potential, before we perform more sophisticated calculations using computational chemistry software.

This chapter allows students to reproduce popular diagrams in physical chemistry and inorganic chemistry freshmen texts using the software the students know by that time. We find the Slater rules – even though nearly forgotten – an interesting approach to develop understanding of the screening concept, which is used for all discussions that involve only valence electrons.

Chapters 7 and 8 are devoted to Hückel theory. We use Hückel theory to motivate the linear combination of atomic orbitals (LCAO) ansatz, which is quite important for any chemical discussion of the electronic properties of molecules. In this book Hückel theory is discussed well beyond the typical *n*-annulenes and linear chain examples. We use this method to show how one can solve the approximate Schrödinger equation within the LCAO concept for quite large systems with no more mathematical tools other than a program to diagonalize a matrix. We apply Hückel theory to determine fullerene stabilities, a method that was even used in "professional computational chemistry" up to the late 1990s. While Chapter 7 is focused on stabilities and the LCAO concept, Chapter 8 discusses bond order, the quantum chemical prediction about the strength of individual bonds, and the resulting molecular orbitals.

Hückel theory is a very intuitive way to teach LCAO, a method that governs quantum chemistry today. Our students found it appealing to work with more "fancy" systems such as fullerenes in addition to the standard ones found in the primary literature. Chapters 7 and 8 form a unit. If there is enough time, they can be performed together. If there is no time, Chapter 8 might be dropped.

While Chapters 2 to 8 have been taught to second-year students, **Chapter 9** is the first of a series that we have designed for third-year students. Chapter 9 provides the basis for geometry optimization. Whereas we cannot discuss the rather complex mathematical apparatus of the large variety of modern algorithms to optimize molecular geometries, we can explain the principal functioning of the steepest descent optimizer and of the Hessian-based Newton–Raphson optimizer on the example of a diatomic. Both algorithms are programmed in a spreadsheet and applied to a diatomic with the PES given by a Morse potential. As more sophisticated algorithms employ the same or similar approximations, this chapter explains why it is important to choose good starting geometries, and why geometry optimizations may fail to find an equilibrium structure. Besides, the typical way to characterize an equilibrium structure by the first and second derivatives of the PES with respect to the coordinates is motivated.

We strongly recommend Chapter 9 to everybody who is going to perform a geometry optimization. It develops the students' sense for iterative algorithms; in principle, they program a geometry optimizer in a spreadsheet that does – essentially – the same as the advanced computer code they are using later. If the student is familiar with a programming language or a math processor the chapter could be adopted to use this knowledge, as a much more useful tool may result from performing the tasks of Chapter 9.

In **Chapter 10** molecules with unpaired spins are treated. Quantities such as the electron spin and multiplicity are introduced. It is shown that it is not always straightforward to predict the spin state of a molecule, and how it can be determined. We further discuss the Jahn–Teller distortion.

This chapter is necessary if systems with unpaired electrons are investigated. You may use this chapter together with Chapter 13 in a course for more advanced students.

Chapters 11 and 12 introduce vibrational spectroscopy of polyatomic molecules. They are based on Chapters 3 and 4, but extend the approach to large systems. In Chapter 11 a simplified calculation of vibrational frequencies and normal modes is given for the carbon dioxide molecule. The assignment of external degrees of freedom to translation and rotation is also discussed. The comparison with IR and Raman spectroscopy experiments is motivated, and the means of coupling of molecular vibrations and light is briefly discussed. Chapter 12 is more advanced. It gives, in a mathematically somewhat more involved way, the calculation of the vibrational spectrum as a system of un-coupled harmonic oscillators and shows that molecular vibrations are indeed independent. More importantly, it relates the molecular vibra-tions to the symmetry of the molecule, and shows how vibrational frequencies can be analyzed in terms of normal modes.

Chapter 11 is designed to introduce vibrational spectroscopy to undergraduates. It might be combined with Chapters 3 and 4. Chapter 12 is definitely beyond the level of a freshmen lab course. It should be given to very interested students, and may accompany a course for higher semesters in computational, analytical or physical chemistry.

Chapter 13 builds on Chapter 6 and shows how ionization potentials and electron affinities can be calculated for molecules. Adiabatic and vertical values (adiabatic and vertical ionization potential and electron affinity as well as vertical detachment energy) are discussed, and their importance for the detection of species with short lifetimes is motivated.

Chapter 13 is a simple lab session that practices the topic of ionization potentials and electron affinities. It is useful as a lab course as well as a hands-on task for a homework accompanying a lecture.

Chapter 14 discusses thermochemistry. The entropy and energy contributions due to translation, rotation, and vibration of the molecule are calculated and used to calculate the thermodynamic properties of the molecule. We also compare relative energies of molecules where the thermodynamic distribution of isomers changes with the temperature.

This chapter looks – at first glance – very difficult. We have used it for third-year students, and, after the initial shock because of the cumber-some mathematical formalism, they really enjoyed it, as the bridge from the single-molecule calculation to statistical thermodynamics is indeed one of the more difficult topics for an undergraduate chemistry student.

Chapters 15 to 17 are devoted to molecular dynamics (MD). We used a combination of these chapters for fourth-year students, but the content is simple enough to be taught to undergraduates. In Chapter 15, the Verlet and velocity Verlet algorithms are introduced, and an MD simulation of diatomic molecules is carried out in a spreadsheet. Emphasis is given to the numerical pitfalls of the MD approach. Chapter 16 introduces thermodynamic ensembles – microcanonical (constant particle number, volume, and energy; NVE) and canonical (constant particle number, volume, and temperature; NVT) – and discusses averaging procedures. It further introduces MD simulations for larger molecules. In Chapter 17 we introduce the simulated annealing approach to search for isomers and the global minimum of molecules and to overcome transition states.

Molecular dynamics is becoming more and more popular in computational chemistry. In a similar way as in Chapter 9 we write a simple MD program in a spreadsheet. Again, this series is easily adoptable to a more advanced programming tool. We use Born–Oppenheimer molecular dynamics, as Chapter 17 (simulated annealing) requires a quantum method for the calculation of forces.

1.6
The Software

This book is accompanied by a CD, which contains a `Linux` operating system based on the `Knoppix` Live system. The CD contains, besides the operating system, a software collection that allows to carry out the tasks in this book to be carried out. The software can be transferred to any computer with a `Linux` system. Most of the programs on the CD, however, can be obtained free of charge for academic usage. The CD compilation on the CD is freeware. You can freely use and redistribute it. We also welcome improvements and further developments. Note that some of the software, in particular the `deMon` computer code, the `CaGe` program, and the `Molden` molecular editor, have different license agreements. They can be used, freely redistributed with this CD or with improvements of the CD project. However, their modification has to be agreed with the original authors.

2
Molecular Coordinates and Symmetry

2.1
Aim

For nearly all tasks in computational chemistry, the molecular coordinates of the investigated molecule, that is, the kind and the position of its atoms, are needed. This introductory chapter describes various ways to define molecular coordinates. The concept of molecular symmetry is introduced and used to reduce the number of relevant structural parameters that are needed to completely characterize a molecule. Furthermore, this chapter is used to make you familiar with frequently applied software such as a molecule viewer and a molecular editor.

molecular coordinates

molecular editor

2.2
Theoretical Background

2.2.1
Cartesian and Internal Coordinates

The position of an atom in space is given by its Cartesian coordinates X, Y, and Z in a three-dimensional Cartesian coordinate system.[1] For a molecular system consisting of N atoms, this leads to $3N$ Cartesian coordinates. As an example, a set of Cartesian coordinates for the water molecule is given in Table 2.1.

For an individual molecule, location and orientation in space are irrelevant, as long as there is no interaction with, for example, an external field. This information (location and orientation) is, however, present in the absolute Cartesian coordinates. Removal of translational

Cartesian coordinates

---

[1] In a three-dimensional Cartesian coordinate system, the axes are perpendicular to each other and the X, Y, and Z axes are aligned following the right-hand rule: the thumb points into X, the index into Y, and the middle finger into Z direction.

Computational Chemistry Workbook: Learning Through Examples
Thomas Heine, Jan-Ole Joswig, and Achim Gelessus
Copyright © 2009 WILEY-VCH Verlag GmbH & Co. KGaA, Weinheim
ISBN: 978-3-527-32442-2

Table 2.1 Cartesian coordinates of the water molecule.

Element	X (Å)	Y (Å)	Z (Å)
O	0.000	0.000	0.000
H	0.959	0.000	0.000
H	−0.230	0.931	0.000

internal coordinates

and rotational degrees of freedom leads to $3N - 6$ internal coordinates for nonlinear molecules, and $3N - 5$ internal coordinates for linear molecules. These internal coordinates describe the relative location of the atoms with respect to each other. For example, one internal set of coordinates for the water molecule ($N = 3$), which has $3N - 6 = 3$ internal degrees of freedom, are the two OH bond lengths and the HOH angle. However, except for the very simple case of a diatomic molecule, internal coordinates are not uniquely defined. Even for a triatomic molecule such as water there are several equivalent sets of internal coordinates (Figure 2.1).

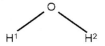

1. OH^1 distance, OH^2 distance, H^1OH^2 angle
2. OH^1 distance, OH^2 distance, H^1H^2 distance
3. H^1H^2 distance, OH^1H^2 angle, OH^2H^1 angle

Figure 2.1 The water molecule and three different sets of internal coordinates to describe it. Any other possible geometric construction of a triangle can be used as well.

It is obvious that for large molecules the number of possible definitions by internal coordinates increases dramatically. A more systematic approach for the definition of a molecular structure by internal coordinates is the so-called Z-matrix method.

2.2.2
The Z-Matrix

Z-matrix

The Z-matrix is – simply speaking – a list of internal coordinates that uniquely describes a molecule. The atoms of the molecule are successively added to the list, and each atom is set in relation to atoms that are already present in the list.

The position of an atom in space is uniquely described by three Cartesian coordinates – as we have seen above – or by three internal coordinates. Since the chemist has the possibility to measure bond lengths, bond angles, or dihedral angles, these are usually used as internal coordinates. But also distances between non-bonded atoms can be used. They are, however, not intuitively understandable.

Table 2.2 Two different Z-matrix descriptions of the water molecule.

No.	Element	A	Bond length (Å)	B	Bond angle (degrees)
1	O				
2	H	1	0.959		
3	H	1	0.959	2	103.9
1	H				
2	O	1	0.959		
3	H	2	0.959	1	103.9

The first atom of the Z-matrix[2] does not have any (relative) coordinates, and consequently no entry in the Z-matrix. The position of the second atom is uniquely defined by stating the bond length between atoms 1 and 2. When adding a third atom, we will have to give a bond length to one of the present atoms, say atom 2, and the bond angle between the two bonds, with previously defined atom 2 being in the apex. The water molecule can therefore be described by different Z-matrices; two possibilities are given in Table 2.2.

The second Z-matrix in Table 2.2 can be described in words as follows. Atom 1 is a hydrogen atom. Atom 2 is an oxygen atom, which is connected to atom 1 (column A states the connectivity for the bonds, that is, the number of the atom to which a particular atom is connected) at a distance of 0.959 Å. Atom 3 is a hydrogen atom, which is connected to atom 2 at the same distance. Moreover, the bond between these two atoms (atoms 3 and 2) has an angle of 103.9° to the bond between atoms 1 and 2 (column B states the respective atom for that angle). Further atoms will also need a statement of the dihedral angle to be uniquely defined.

To summarize, each atom, except for the first three, is defined by three internal coordinates: a bond length, a bond angle, and a dihedral angle, with respect to three reference atoms (A, B, C). The reference atoms must already have been defined, and therefore their reference numbers are lower than the number of the newly added atom. **reference atom**

2.2.3
The Dihedral Angle

The bond length and the bond angle can be understood intuitively. The dihedral angle sometimes causes problems, because its definition uses the normal vectors of two planes. Each of the planes is defined by three **bond length** **bond angle** **dihedral angle**

[2] The name is given following a convention to put the first atom in the origin and the second on the Cartesian Z axis of the system.

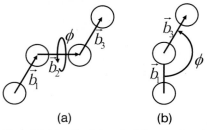

Figure 2.2 The dihedral angle defined by three bonds between four atoms: (a) side view; (b) view along the $\boldsymbol{b}_2$ vector. Here, the marked dihedral angle ϕ has a positive value.

atoms, and the resulting normal vector generally does not point along a chemical bond.

There is, however, a more chemical way of finding the dihedral angle, **Newman projection** which is used in organic chemistry, the so-called Newman projection. Since the dihedral angle is defined by four atoms, we take four atoms that are – if possible – connected in a row (Figure 2.2a). This is not mandatory, but helpful. These four atoms define three bonds (in Figure 2.2a these are the vectors $\boldsymbol{b}_1$, $\boldsymbol{b}_2$, and $\boldsymbol{b}_3$). The molecule is then rotated in such a way that the vector $\boldsymbol{b}_2$ is pointing towards us (Figure 2.2b). Now the dihedral angle is defined as the angle between the other two remaining vectors $\boldsymbol{b}_1$ and $\boldsymbol{b}_3$. It is, thus, an angle between two unconnected bonds.

A clockwise rotation of vector $\boldsymbol{b}_3$ with respect to vector $\boldsymbol{b}_1$ is marked with a plus sign, a counterclockwise rotation with a minus sign. The **torsion angle** dihedral angle is also called the torsion angle. In order to find a dihedral angle in a molecule, it needs to have at least four atoms. Tables 2.3 and 2.4 show two definitions of the methane molecule by either Cartesian coordinates or internal coordinates (Z-matrix).

An intuitive way to select the reference atoms is to follow the chemical bonding of the molecule. This is not mandatory, and reference atoms that are not connected by a chemical bond can also be used. But bond lengths and bond angles are usually known, and therefore it is helpful to use them. Defining the Z-matrix alongside the molecule's bonding pattern is straightforward.

Table 2.3 Cartesian coordinates of the methane molecule.

Element	X (Å)	Y (Å)	Z (Å)
C	0.000	0.000	0.000
H	0.000	0.000	1.089
H	1.027	0.000	−0.363
H	−0.513	−0.889	−0.363
H	−0.513	−0.889	−0.363

Table 2.4 Internal coordinates of the methane molecule, written in the Z-matrix representation.

	Element	A	Bond length (Å)	B	Bond angle (degrees)	C	Dihedral angle (degrees)
1	C						
2	H	1	1.089				
3	H	1	1.089	2	109.471		
4	H	1	1.089	2	109.471	3	120.000
5	H	1	1.089	2	109.471	3	−120.000

Despite the advantage that relevant data (bond lengths and bond angles) can be read directly from the Z-matrix, the Z-matrix representation also has clear disadvantages. First, more information is needed, as the reference atom and variables have to be passed simultaneously. Then, there is the problem of numerical processing: bond angles and dihedral angles require the use of trigonometric functions, which are not linear. This may have consequences on the numerical precision.

Many calculations in computational chemistry include an optimization of the molecule's geometry with respect to the used potential. Therefore, the exact values for the molecular coordinates are not needed, or even may be completely unknown. It is only necessary to give a rough estimate of the bonding in the molecule, in order to start the calculation. To start a quantum chemical calculation, only the topology and a rough estimate of the bonding parameters are necessary. For their prediction, chemical intuition and knowledge of average bond lengths are useful.

2.2.4
Symmetry

From everyday life we already know about the concept of symmetry. The most popular symmetry operation is probably mirroring. If a molecule can be transformed by a mathematical operation into an image that corresponds to the initial structure, we call this operation a symmetry **symmetry operation** operation. For example, rotation by 360° around any axis is a symmetry operation. The respective axis is called a symmetry element (of the **symmetry element** molecule). Besides rotation, other symmetry operations are mirroring and inversion. Performing no symmetry operation is equivalent to the trivial symmetry operation, the rotation by 360°, and is called identity. It is considered to be a symmetry operation itself. Symmetry elements are rotational axes (stating by how many degrees a rotation should be performed), mirror planes, and the center of inversion. There are also higher-order symmetry elements and operations that combine these basic operations.

A molecule that has one or more non-trivial symmetry elements is called symmetric. The advantage of the presence of symmetry elements is that they reduce the number of internal coordinates needed to describe the molecular structure. For example, we know that in the water molecule both OH bonds have the same length (the water molecule therefore has a mirror plane perpendicular to the molecular plane). Therefore, only two internal coordinates – the OH distance and the H^1OH^2 angle – are sufficient to fully define the structure, because we know that the two OH bonds are equal in length. Any other two internal coordinates, for example, the OH distance and the H^1H^2 distance, could also be used.

For the highly symmetric methane molecule, the number of internal coordinates can be reduced even further, from nine to one, using symmetry constraints. Since we know that the methane molecule has a tetrahedral arrangement of the hydrogen atoms, the only freely adjustable parameter for methane is the CH bond length (or, equivalently, the HH distance). A thorough discussion of symmetry concepts and their application to chemistry requires group theory, which is explained in detail in many textbooks [1–3]. Here, a brief introduction to the topic is presented, which should suffice for a basic understanding.

2.2.5
Symmetry Elements and Operations

The examples of water and methane have already shown that molecules can possess symmetry elements. First, we will discuss briefly the symmetry of the water molecule (Figure 2.3). The molecular plane

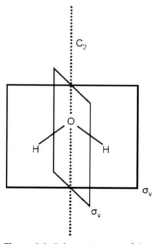

Figure 2.3 Schematic view of the water molecule and its symmetry elements: a rotational C_2 axis and two vertical σ_v mirror planes. The identity operation usually is not shown.

defined by the three atoms is at the same time a mirror plane (also called reflection plane). We label this symmetry element with the symbol σ and is denoted σ. Upon execution of the reflection operation at this plane, each atom remains in its original position and the resulting image corresponds to the initial structure. The water molecule possesses a second mirror plane σ. It is perpendicular to the first one, passes through the oxygen atom, and bisects the H^1OH^2 angle. Reflection at this mirror plane leaves the oxygen atoms in its place and interchanges the two hydrogen atoms. The resulting image is nevertheless indistinguishable from the initial structure.

mirror plane

Besides the two mirror planes, the water molecule also has a rotational axis, which passes through the oxygen atom and bisects the H^1OH^2 angle. This axis is, moreover, the intersection of the two mirror planes. Rotating the molecule around this axis by 180° interchanges the two hydrogen atoms, but the image is exactly on top of the initial structure. If the rotation is applied twice, all the atoms are back to their original position. Therefore, this rotational axis is called a 2-fold axis and is denoted C_2. If n rotations (each by an angle of $360°/n$) are necessary to obtain congruence, a rotational axis is called an n-fold axis and is denoted C_n.

rotational axis

The suffix notation "v" for the mirror plane stands for vertical mirror plane (σ_v in Figure 2.3). This means that the highest rotational axis (here, the C_2 axis) lies in the mirror plane. The highest rotational axis is that n-fold axis with the highest n, that is, the smallest rotation angle. There is also the possibility of the highest rotational axis being perpendicular to the plane, in which case the plane is called horizontal mirror plane (σ_h).

vertical mirror plane

horizontal mirror plane

Finally, there is a fourth, the trivial symmetry element. Under the unity operation (identity) E, all the atoms remain in their original positions. Other possible symmetry elements for molecules are a rotation–reflection axis S_n (also called improper rotation axis or alternating axis) and a center of inversion, called i.

rotation-reflection axis
center of inversion

2.2.6
Point Groups

The set of symmetry operations for a molecule fulfills the mathematical definition of a group. Hence, the principles of group theory can be applied to symmetric molecules. The symmetry groups are called point groups and labeled according to their symmetry elements. The identity (unity operation) E is element of all point groups. The following 10 point group families exist in chemistry.

point groups

- Non-axial groups: groups without any rotational axis, C_1, C_s, C_i.
- C_n groups: only an n-fold rotational axis C_n exists.

- D_n groups: an n-fold rotational axis C_n plus n C_2 axes perpendicular to C_n exist.
- C_{nv} groups: C_n group plus n vertical mirror planes incorporating C_n.
- C_{nh} groups: C_n group plus a mirror plane perpendicular to axis C_n.
- D_{nh} groups: D_n group plus a mirror plane perpendicular to axis C_n.
- D_{nd} groups: D_n group plus n mirror planes incorporating C_n.
- S_n groups: only an n-fold rotation–reflection axis exists (n must be even).
- Cubic groups: seven tetrahedral (T, T_h, T_d), octahedral (O, O_h), and icosahedral (I, I_h) groups for highly symmetrical molecules; there must be more than one rotational axis C_n with $n \geq 3$.
- Linear groups: groups $C_{\infty v}$ and $D_{\infty h}$ for linear molecules.

A scheme to find the point group of a molecule by simply checking for existing elements is given in Figure 2.4. According to this scheme (and

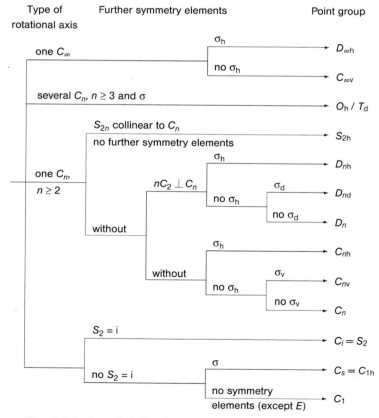

Figure 2.4 A scheme for finding the point group of a molecule. Start at the left-hand side and check for required symmetry elements.

the definitions above) the water molecule H_2O belongs to the point group C_{2v}.

As discussed above, the number of coordinates necessary to define a molecular structure can be reduced using internal coordinates and symmetry properties. On the other hand, the development of algorithms for computational chemistry is more straightforward if Cartesian coordinates are used. Many computational chemistry computer codes first transform internal coordinates to Cartesian coordinates and perform the calculation using these. If necessary, the final results are transformed back to internal coordinates. For large molecules the use of internal coordinates is less appealing, because the definition becomes very cumbersome and the removal of six degrees of freedom does not reduce the number of coordinates significantly. In addition, symmetry properties are usually of less importance for large molecules; the fullerenes are a famous, but rare, exception to this rule.

2.3
Demonstration

2.3.1
Constructing a Z-Matrix

In the following, we will build up a Z-matrix for the planar formamide molecule ($HCONH_2$). The structure is depicted in Figure 2.5a, in which the atoms are already labeled. The bond lengths and angles will be given below.

Defining the internal coordinates for the first three atoms is straightforward. We start with the C^1 atom, and add the N^2 atom at a distance of R_{CN}. Then the O^3 atom is added, connected to atom C^1 (at a distance R_{CO}) with an NCO angle α_{NCO}. The Z-matrix then reads as:

1	C				
2	N	1	R_{CN}		
3	O	1	R_{CO}	2	α_{NCO}

The hydrogen atoms numbered 4 and 5 are bonded to the nitrogen atom (amino group). Since the molecule is planar (all atoms lie in the same plane), there are only two possibilities for the dihedral angle (see Figure 2.5b): H^4 has a 180° dihedral angle, and H^5 a 0° dihedral angle (both dihedrals are relative to the atoms 1, 2, and 3). The last atom, H^6, is bonded to the central carbon atom and has a 180° dihedral angle (relative to atoms 1, 2, and 5). The total Z-matrix, therefore, reads as:

1	C						
2	N	1	R_{CN}				
3	O	1	R_{CO}	2	α_{NCO}		
4	H	2	R_{NH}	1	α_{H4NC}	3	ϕ_{H4NCO}
5	H	2	R_{NH}	1	α_{H4NC}	3	ϕ_{H5NCO}
6	H	1	R_{CH}	2	α_{H6CN}	5	ϕ_{H6CNH5}

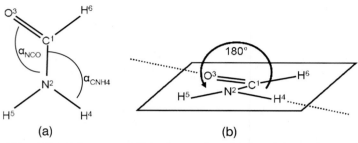

Figure 2.5 (a) The structure of the formamide molecule and the numbering of the atoms; (b) schematic view along the N^2C^1 axis to see the $O^3C^1N^2H^4$ dihedral angle (180°).

The bonding parameters are as follows: $R_{CN} = 1.368$ Å, $R_{CO} = 1.212$ Å, $R_{NH} = 1.027$ Å, $R_{CH} = 1.125$ Å, $\alpha_{NCO} = 125.0°$, $\alpha_{H4NC} = 119.2°$, $\alpha_{H6CN} = 120.0°$. The dihedral angles can be obtained from the information that the molecule is planar (Figure 2.5b).

Next, we will search for symmetry elements and the point group. A planar molecule has at least one mirror plane, the molecular plane (Figure 2.5b). In fact, we find that this is the only symmetry element for the formamide molecule. There are no more mirror planes, no rotational axes, and no center of inversion. According to the scheme in Figure 2.4, the molecule therefore has C_s symmetry.

2.3.2
Determining Molecular Point Groups

In the second part of this Demonstration, as an example, we will search for symmetry elements in selected molecules and determine their point groups. We start with the carbon monoxide (CO) and carbon dioxide (CO_2) molecules (Figure 2.6). Both molecules are linear, and it is possible to rotate the molecules around their molecular axes by any angle resulting in an indistinguishable image. The smallest possible angle is an infinitesimally small angle. Therefore, these axes are called C_∞ axes. The scheme in Figure 2.4 shows us that there are only two point groups in which a C_∞ axis is present, $C_{\infty v}$ and $D_{\infty h}$. They differ by the presence of a horizontal mirror plane, a mirror plane perpendicular to this highest rotational axis.

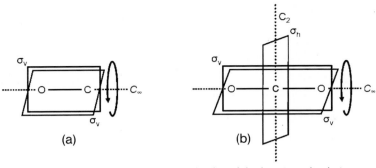

Figure 2.6 Schematic view of (a) the CO molecule and (b) the CO_2 molecule (no chemical bonding is indicated). Some symmetry elements are shown.

Both molecules have, moreover, an infinite number of σ_v planes. Two of them are indicated in each molecule in Figure 2.6. For the CO molecule we do not find any mirror plane perpendicular to the main (and only) rotational C_∞ axis. Therefore, this molecule has $C_{\infty v}$ symmetry. For the CO_2 molecule the situation is different. The molecule is symmetric with respect to the central O atom, and we find an infinite number of C_2 axes perpendicular to the main axis (C_∞). One of them is indicated. All these C_2 axes lie within a horizontal σ_h plane. At the intersection of all the axes, which is the position of the O atom, the molecule has additionally a center of inversion (not indicated). Besides the C_∞ axis, the CO_2 molecule has a σ_h plane. These two symmetry elements are enough to find the point group: $D_{\infty h}$.

Next, we will determine the point group of the ammonia molecule, NH_3 (Figure 2.7). It is a non-planar molecule with a C_3 axis (through the nitrogen atom). There are, moreover, three σ_v planes. Using the scheme in Figure 2.4, we find the point group to be C_{3v}.

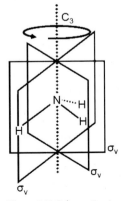

Figure 2.7 Schematic view of the NH_3 molecule and its symmetry elements.

2.4
Problems

1. Molecular coordinates

Define the structures of the following molecules by either Cartesian coordinates or the Z-matrix.

a) Water (H_2O): $R_{OH} = 0.9575$ Å, $\alpha_{HOH} = 104.51°$.

b) Formic acid (HCOOH, planar): $R_{CH} = 1.097$ Å, $R_{CO} = 1.343$ Å (hydroxyl), $R_{CO} = 1.202$ Å (carbonyl), $R_{OH} = 0.972$ Å, $\alpha_{HCO} = 124.1°$ (carbonyl), $\alpha_{OCO} = 124.9°$, $\alpha_{COH} = 106.3°$.

c) 1,2,5-Oxadiazole ($C_2H_2N_2O$, planar): $R_{ON} = 1.380$ Å, $R_{CN} = 1.300$ Å, $R_{CC} = 1.421$ Å, $R_{CH} = 1.076$ Å, $\alpha_{NON} = 110.4°$, $\alpha_{ONC} = 105.8°$, $\alpha_{CCN} = 109.0°$, $\alpha_{CCH} = 130.2°$, $\alpha_{NCH} = 120.9°$.

2. Construction of molecular structures

Construct the structures of acetone, borazine and benzene using a molecular editor.

3. Molecular structure and symmetry

Determine the symmetry elements and point groups for the six molecules used in problems 1 and 2. If you are familiar with group theory, you can do this by hand. If not, you can get help from the program "Character Tables for Chemically Important Point Groups" (see Appendix).

4. Internal coordinates and structure

Determine the number of internal coordinates for borazine, if symmetry constraints are applied. Again, you can get help from the program "Character Tables for Chemically Important Point Groups" to answer this question. Suggest two sets with the minimum number of internal coordinates for borazine.

Hints

1. The structures of 1,2,5-oxadiazole, borazine and benzene are:

2. Verify your results from problem 1 with a molecular viewer. Write your coordinates (internal or Cartesian) into a file and load this into the molecular viewer.

2.5
Technical Details

1. Throughout this experiment you can use the combined molecular editor and viewer `Molden`, which you can find on the DVD distributed with this book.
2. For problem 2, use the function **Substitute atom by fragment** in the Z-matrix editor of `Molden`.

2.6
Review and Summary

We have introduced Cartesian coordinates, internal coordinates, and the Z-matrix concept. Whereas Cartesian coordinates give an absolute description of the molecule in space, the internal coordinates in terms of the Z-matrix define the relative location of the atoms as well as the orientation of bonds with respect to each other. The atomic positions are uniquely defined by a bond length, an angle, and a dihedral angle. Following the chemical bonding pattern is very helpful, to build a Z-matrix.

We have also discussed the concept of symmetry and introduced several, though not all, symmetry elements: rotational axes, vertical and horizontal mirror planes, and the center of inversion. To use symmetry elements with group theory, an identity element is necessary; the corresponding symmetry operation leaves all atoms in their original places.

Searching for the symmetry elements of a molecule can lead to the determination of the molecule's point group. A scheme (Figure 2.4) helps in the determination of the right point group with the help of only a few relevant symmetry elements.

References

1 Cotton, F.A. (1990) *Chemical Applications of Group Theory*, 3rd edn, John Wiley & Sons, Inc.

2 Bishop, D.M. (1993) *Group Theory and Chemistry*, Dover Publications.

3 Atkins, P.W. and de Paula, J. (2006) *Physical Chemistry*, 4th edn, Oxford University Press.

3
Vibrations of Diatomic Molecules: The Harmonic Approximation

3.1
Aim

We calculate the vibrational frequency of diatomic molecules within classical theory using analytical mechanics. In this experiment, you learn to work with one of the most important approximations in physics as well as in computational chemistry, the harmonic approximation, which will be used later in this book for geometry optimization, the vibrational analysis of large molecules, and thermochemistry. We introduce the Morse potential, a good approximation for the potential energy surface of diatomic molecules. You learn to formulate the physical problem for the calculation of vibrational frequencies. In the next chapter, we will extend this approach to quantum mechanics and introduce the Schrödinger equation. This experiment is designed for students without higher mathematical background, and can be skipped if its content is known.

harmonic approximation

3.2
Theoretical Background

3.2.1
The Harmonic Approximation

It is well known from experiments measuring the vibrational spectrum of molecules that the potential energy surface of diatomic molecules can be reasonably well modeled using an empirical two-body potential, the Morse potential:

potential energy surface
Morse potential

$$V(R) = D_e \left\{ [1 - \exp(-\alpha(R - R_e))]^2 - 1 \right\} \tag{3.1}$$

Here, R is the distance between the two nuclei (see Figure 3.1). All the parameters of Equation 3.1 can be interpreted in chemical terms: $-D_e$ is the value of the absolute minimum of the function at the equilibrium

Computational Chemistry Workbook: Learning Through Examples
Thomas Heine, Jan-Ole Joswig, and Achim Gelessus
Copyright © 2009 WILEY-VCH Verlag GmbH & Co. KGaA, Weinheim
ISBN: 978-3-527-32442-2

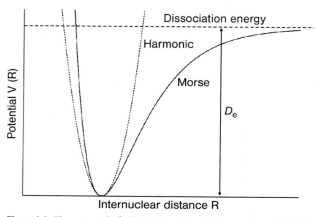

Figure 3.1 The potential of a diatomic molecule (Morse potential, solid line) and its harmonic approximation (dotted line).

position R_e and is associated with the binding energy of the molecule, and α is a constant that is related to the curvature of the function close to the equilibrium position. For short distances, the potential becomes very repulsive; for large distances, the potential describes dissociation, where the binding energy and consequently the potential approach zero. You will use this or very similar potential functions whenever you tackle a problem using a classical force field, which are used frequently in physics, chemistry, and biology (for further reading see [1]).

Like most other systems on Earth, molecules spend their time mostly oscillating around their equilibrium positions, except that they just happen to participate in a chemical reaction or they are permanently in an excited state, for example, in a plasma. Therefore, as long as no bond breaking or bond formation is in progress, it is for most purposes sufficient to know the potential function only close to the equilibrium position. This picture changes only if there is enough energy available to activate high-energy oscillations or even a chemical reaction. For a diatomic molecule in the gas phase, this means that the energy is in the order of magnitude of D_e. Independently of how complicated the exact form of the potential is, we can approximate it well around the equilibrium position R_e by expanding it in a Taylor series:

$$V(R) \approx V(R_e) + \underbrace{\left.\frac{\partial V(R)}{\partial R}\right|_{R=R_e}}_{0} (R-R_e) + \frac{1}{2}\left.\frac{\partial^2 V(R)}{\partial R^2}\right|_{R=R_e} (R-R_e)^2 + \cdots \tag{3.2}$$

As the potential has a minimum at R_e, its first derivative with respect to R vanishes. And if third- and higher-order terms are neglected, the potential has the form of a parabola:

$$V(R) \approx V(R_e) + \frac{1}{2}\frac{\partial^2 V}{\partial R^2}\bigg|_{R=R_e} (R-R_e)^2 \tag{3.3}$$

Now we subtract $V(R_e)$ from both sides and obtain the harmonic **harmonic potential** potential, a very simple approximation to the potential at the equilibrium position, depending on the single parameter $\partial^2 V/\partial R^2|_{R=R_e}$:

$$V_{\mathrm{harm}}(R) = \frac{1}{2}\frac{\partial^2 V}{\partial R^2}\bigg|_{R=R_e} (R-R_e)^2 \tag{3.4}$$

In practice, we can work with Equation 3.4, as a constant shift of the potential will not affect the laws of physics. Note that Equation 3.4 holds independently of the potential form, which allows its application even if the potential is only known numerically. We would like to determine the frequency of the molecular vibration; therefore, we must find the equations of motion of the atoms. This can be done by calculating the force F acting on the atoms with interatomic distance R, which is the negative gradient of the potential at the same position. Within the harmonic approximation, we obtain a linear relation between F and R. This is also known as Hooke's law:

$$F(R) = -\frac{\partial V_{\mathrm{harm}}(R)}{\partial R} = -\frac{\partial^2 V}{\partial R^2}\bigg|_{R=R_e} (R-R_e) = K(R-R_e) \tag{3.5}$$

We take advantage of the fact that $-\partial^2 V/\partial R^2|_{R=R_e} = K$ is constant and call this quantity the force constant. Moreover, we know from Newton's **force constant** law that the force F equals mass times acceleration:

$$F(R) = K(R-R_e) = \mu\frac{\partial^2 R}{\partial t^2} \tag{3.6}$$

The acceleration is obviously the second derivative of R with respect to time, while the meaning of the mass μ in Equation 3.6 is not obvious.

3.2.2
The Reduced Mass

We need to keep in mind that we want to describe the motion of two atoms (1 and 2) around their equilibrium positions R_{e1} and R_{e2} (for further reading see [2]). If we assume that the X axis of the coordinate system is aligned with the molecular bond, the atoms take positions X_1 and X_2 with the bond length $R = X_2 - X_1$ (the Y and Z coordinates are zero, and we assume that $X_2 > X_1$). The force that each atom experiences is equal in magnitude, but opposite in sign:

$F = F_2 = -F_1$. Applied to the two individual nuclei, we obtain

$$m_1 \frac{\partial^2 X_1}{\partial t^2} = F, \quad m_2 \frac{\partial^2 X_2}{\partial t^2} = -F \;\Rightarrow\; \frac{\partial^2 X_2}{\partial t^2} = -\frac{m_1}{m_2}\frac{\partial^2 X_1}{\partial t^2} \qquad (3.7)$$

and understand easily that the force accelerates light nuclei more strongly than heavier ones. Since we have used the bond length R, which can be considered as a relative coordinate, we need a relative acceleration as well:

$$\frac{\partial^2 R}{\partial t^2} = \frac{\partial^2 X_1}{\partial t^2} - \frac{\partial^2 X_2}{\partial t^2} = \left(1 + \frac{m_1}{m_2}\right)\frac{\partial^2 X_1}{\partial t^2} = \frac{m_2 + m_1}{m_2}\frac{\partial^2 X_1}{\partial t^2} \qquad (3.8)$$

Here, we make use of the right-hand side of Equation 3.7. Moreover, we have written Newton's law in Equation 3.6 using the mass μ. To calculate μ we now use Equation 3.8:

$$\mu = \frac{F}{\partial^2 R/\partial t^2} = \frac{F}{\partial^2 X_1/\partial t^2}\frac{m_2}{m_2 + m_1} = m_1\frac{m_2}{m_2 + m_1}$$

reduced mass This is the definition of the so-called reduced mass μ (note that it does have the unit of mass). It can also be computed using the more intuitive relation

$$\frac{1}{\mu} = \frac{1}{m_1} + \frac{1}{m_2} \qquad (3.9)$$

3.2.3
The Vibrational Frequencies

As next step, we need to find the function that relates the bond length with time, $R(t)$, which can be obtained by solving the differential Equation 3.6. It is useful to imagine the new setup of the vibrating molecule, as indicated in Figure 3.2. The vibration is described by a fictitious particle of mass μ that is connected by a spring to a fixed wall. Let us introduce the more convenient variable $\tilde{R}(t) = R(t) - R_e$, giving the elongation of the molecule from its equilibrium position. $\tilde{R}(t)$ becomes positive if the bond length is stretched and negative if it is compressed. Equation 3.6 becomes (as R_e is fixed)

$$\frac{\partial^2 \tilde{R}}{\partial t^2} - \frac{K}{\mu}\tilde{R} = 0 \qquad (3.10)$$

Thus, we look for a function whose second derivative gives a function times a negative constant. An obvious solution is the sine function,

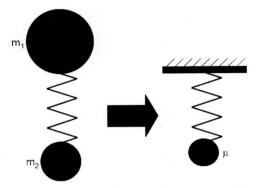

Figure 3.2 Model transformation by introduction of the reduced mass. The left-hand side shows two masses connected by a spring. The right-hand side shows the transformed model system, where the reduced mass μ is fixed to a wall using the same spring.

$$\tilde{R}(t) = A \sin(\omega_0 t) \tag{3.11}$$

whose second time derivative is

$$\frac{\partial^2 \tilde{R}(t)}{\partial t^2} = -A\omega_0^2 \sin(\omega_0 t)$$

Here A determines the amplitude of the oscillation, and $\omega_0 = 2\pi f_0$ its angular frequency, which is directly related to the vibrational frequency f_0, but is more convenient to use as it avoids having to carry the factor 2π in the argument of the sine function. Expressing $\tilde{R}$ and its second derivative in Equation 3.11 by the sine function, we end up with

vibrational frequency

$$-A\omega_0^2 \sin(\omega_0 t) - \frac{K}{\mu} A \sin(\omega_0 t) = 0 \implies \omega_0 = \sqrt{-\frac{K}{\mu}} \tag{3.12}$$

The solution for ω_0 is always real, as the reduced mass is always positive, and, as defined, K is negative (cf. Equation 3.6). We have shown now that an oscillating molecule has a characteristic frequency, which can be calculated as

$$f_0 = \frac{\omega_0}{2\pi} = \frac{1}{2\pi} \sqrt{-\frac{K}{\mu}} \tag{3.13}$$

The amplitude of the oscillation of the molecule is arbitrary within classical physics, as indicated in Equation 3.11. It will be shown in the next chapter that this result is an artifact of classical physics and does not hold if we use a quantum mechanical approach.

It remains to change the result for the fictitious particle with mass μ back to the motion of the individual atoms. Following Equation 3.7, and with $R = X_2 - X_1$, we can show easily that the amplitudes also relate as $m_1 A_1 = m_2 A_2$, and the total amplitude must be $A = A_1 + A_2$. Both atoms oscillate with the same frequency. Consequently, we obtain, for each individual atom,

$$A_1 = A\left(1 - \frac{m_1}{m_2}\right), \quad X_1 = X_{e1} + A_1 \sin(\omega_0 t)$$

$$A_2 = A\left(1 - \frac{m_2}{m_1}\right), \quad X_2 = X_{e2} - A_2 \sin(\omega_0 t)$$

(3.14)

The collective motion of the atoms in the molecule is called a vibrational mode. The two atoms oscillate around their equilibrium positions, moving collectively towards and away from each other. Our calculation shows that the vibrational frequency depends only on the masses of the atoms in the molecule and on the second derivative of the potential with respect to the bond length at the equilibrium position (Equations 3.13 and 3.9). Within the classical calculation, the oscillations can take any amplitude A. We will see in the next chapter that classical physics indeed gives the correct values for the vibrational frequency, but it is not true that any value for the amplitude A is possible.

Experimentally, molecular oscillations are measured by light. If atoms move, the molecular dipole moment and the polarizability can changes with the bond length. If this is the case energy can be transferred to or from the molecule and can be measured.

3.3
Demonstration

We investigate the vibrational frequencies of the carbon monoxide (CO) molecule. The Morse potential parameters, as obtained by experiment, are given in Table 3.1. We use the Taylor series expansion of the potential (Equation 3.2) to apply the harmonic approximation to the Morse potential:

$$V(R) \approx V(R_e) + \underbrace{\left.\frac{\partial V(R)}{\partial R}\right|_{R=R_e}}_{0} (R-R_e) + \frac{1}{2}\left.\frac{\partial^2 V(R)}{\partial R^2}\right|_{R=R_e} (R-R_e)^2 + \cdots$$

The first derivative of the potential with respect to the bond length is zero, because we are at the minimum of the potential (at position R_e).

Table 3.1 Morse potential parameters for selected diatomic molecules.

	D_e (kJ mol^{-1})	α (10^8 cm^{-1})	R_e (Å)
CO	1072.8	2.312	1.13
H$_2$	432.3	1.995	0.74
HF	566.2	2.259	0.92
HCl	427.9	1.900	1.27
HBr	362.6	1.843	1.41
HI	294.7	1.786	1.61
F$_2$	155.1	3.022	1.14
Cl$_2$	238.9	2.019	1.99
Br$_2$	189.1	2.007	2.28
I$_2$	147.4	1.876	2.67

Source: *CRC Handbook of Chemistry and Physics*, [3].

Higher-order terms are neglected. Setting the second derivative at R_e equal to K (as above), we end up with the harmonic expression of the potential:

$$V(R) \approx V(R_e) + \frac{1}{2}K(R-R_e)^2$$

Note that we are looking at the potential, not at the harmonic potential V_{harm}. Moreover, K is positive, since we are not looking at the derivative here. Next we need the first and second derivatives of the Morse potential (Equation 3.1), which are

$$\frac{\partial V(R)}{\partial R} = 2D_e\alpha\left[(e^{-\alpha(R-R_e)}) - (e^{-2\alpha(R-R_e)})\right]$$

$$\frac{\partial^2 V(R)}{\partial R^2} = 2D_e\alpha^2\left[(2e^{-2\alpha(R-R_e)}) - (e^{-\alpha(R-R_e)})\right]$$

$$(3.15)$$

At position R_e the exponential functions equal one, so that we end up with

$$\left.\frac{\partial^2 V(R)}{\partial R^2}\right|_{R=R_e} = 2D_e\alpha^2 \Rightarrow K = 2D_e\alpha^2$$

The values for CO in Table 3.1 result in a value of $K_{mol} = 1.147 \times 10^{27}$ kg mol^{-1} s^{-2} (the unit equals J mol^{-1} m^{-2}). This is a molar quantity that we can convert to an absolute energy value by dividing it by Avogadro's number: $K = 1904.37$ kg s^{-2}. Alternatively, we could have converted D_e into an absolute energy value first and then calculate K. To calculate the resulting angular frequency, we need the (reciprocal) reduced mass of CO, which is $1/\mu = 8.779 \times 10^{25} \, \text{kg}^{-1}$. The angular frequency is then

$$\omega_0 = \sqrt{\frac{K}{\mu}} = 4.09 \times 10^{14} \, \text{s}^{-1}$$

Typically, the unit for vibrations is "wavenumbers",

$$\bar{v} = \frac{\omega_0}{2\pi c} = 2170 \, \text{cm}^{-1}$$

3.4
Problems

1. Manual calculation of vibrational frequencies of diatomic molecules

Calculate the vibrational frequencies for hydrogen and the series of hydrogen halide molecules (HX, X = H, F, Cl, Br, I). The Morse potential parameters are given in Table 3.1. Compare the calculated values with experimental values from the literature [3, 4].

2. Computer calculation of vibrational frequencies of diatomic molecules

Repeat the calculations using modern tools of computational chemistry, that is, perform a frequency analysis on the basis of a potential that is obtained by solving the electronic Schrödinger equation of the molecule in a approximate way, for example, at the level of density-functional theory (see Technical Details). Compare the calculated values for the vibrational frequency with the values you computed from experimental data in problem. Note that most computational chemistry codes also make use of the harmonic approximation.

3. Oscillations of diatomic molecules

Draw the two atoms of each molecule HX of problem 1 in their equilibrium structure. Assume that the H atom is oscillating with amplitude $A = 0.25 \, \text{Å}$. Determine and indicate the amplitude of the oscillation of the X atom.

4. Role of the atomic masses on the amplitudes

How would the results of problem 1 change if the force constants of all HX molecules were the same? Draw the vibrational frequencies of the HX molecules as a function of the mass of the halogen atom.

5. The reduced mass

Rationalize the fact that for heavy X the reduced mass becomes close to the mass of a hydrogen atom.

3.5
Technical Details

All the problems except problem 2 can be solved using pen and paper. For problem 2, the deMon input file for the Demonstration example of carbon monoxide looks as follows. The initial bond length of 1.1 Å has been taken from the literature. You can type this input file using any editor, for example Nedit. Save the file as co.inp.

```
TITLE FREQUENCY CO
VXCTYPE PBE
BASIS (DZVP-GGA)
AUXIS (GEN-A2)
OPTIMISATION
FREQUENCY
GEOMETRY CARTESIAN
C  0.0 0.0 0.0
O  0.0 0.0 1.1
END
```

So we request from deMon to solve the Schrödinger equation within a certain approximation, called density-functional theory with gradient corrections (line 2), and within a certain mid-quality basis set (lines 3–4). It is supposed to find the equilibrium structure by optimizing (line 5) the geometry (see Chapter 9 for details) and to perform a vibrational (frequency) analysis (line 6). Line 7 specifies that the geometry is following in Cartesian coordinates in Ångstroms. The keyword END indicates the end of the geometry. After running the calculation (type deMon co), you obtain a couple of output files. First, inspect the file co.out, which contains the most important information. It is, of course, not necessary to study the whole output, but a few messages should be checked. First, we need to see if the geometry optimization has finished successfully. If it has been successful, the output contains the message

```
*** THE GEOMETRY IS OPTIMIZED ***
```

If there are no further error messages, the run was successful, and further information can be found in the output. It is, however, easier to read this information using Molden. For this purpose, a Molden interface file (co.mol) is generated automatically, which can be interfaced with Molden by typing molden co.mol. If the main window remains empty, you need to specify the **Ball & Stick** option under the button **Solid** and possibly to activate the **Shade** option. Then you can rotate and move the molecule using the mouse (check the Molden manual for further instructions). At the top right is a button

Norm. Mode that can be activated. We see now the vibrational mode – of course, only one, as it is a diatomic molecule. Clicking at the frequency (given in cm^{-1}), the normal mode, that is, the vibration of the molecule, is visualized.

3.6
Review and Summary

We have seen that the vibrational frequency of a diatomic molecule is governed by the curvature of its potential energy surface – the plot $V(R)$ as shown in Figure 3.1 – at the equilibrium position, and by the masses of both atoms. The harmonic approximation is a simple, albeit reasonably accurate, method to compute vibrational frequencies. It is most of all convenient, as it only requires knowledge of the curvature of the potential at the equilibrium position. The approach allows vibrational frequencies to be assessed, but fails to explain that nature only allows distinct amplitudes for the vibration. In the next chapter the same problem is treated using quantum mechanics, and these shortcomings will be resolved.

For the calculation of vibrational frequencies, we have combined the harmonic approximation with Newton's second law to find an expression for the force in a vibrating system. Thereby, the reduced mass has been introduced. The problem of two moving nuclei has, thus, been reduced to one mass connected by a spring to a fixed wall. Using the harmonic approximation, we have been able to calculate the vibrational frequency of the carbon monoxide molecule in a classical description.

References

1 Force fields are discussed in great detail in Chapter 2 of Cramer, Christopher C. (2004) *Essentials of Computational Chemistry – Theory and Models*, 2nd edn, John Wiley & Sons, Inc.

2 A solid physical basis on the analytical treatment of two-particle problems is provided in analytical mechanics textbooks, for example, Greiner, W. (2003) *Classical Mechanics – Point Particles and Relativity*, Springer, New York.

3 All data are taken from Lide, D.R. (ed.) (2008) *CRC Handbook of Chemistry and Physics*, 89th edn CRC Press, Boca Raton, FL.

4 A good source of information is the web page of the National Institute of Standards of the USA, NIST (http://webbook.nist.gov/chemistry).

4
Vibrations of Diatomic Molecules:
The Schrödinger Equation

4.1
Aim

In this experiment, we introduce the central equation of quantum chemistry, the stationary Schrödinger equation. In quantum chemistry, the Schrödinger equation is usually solved for the electrons in a molecule. Instead, we start here with a simpler example by treating the vibrations of diatomic molecules. It turns out that the mathematical formulation of our problem is, in good approximation, the quantum mechanical harmonic oscillator, one of the examples that can be found in almost all texts on quantum physics. In this experiment, you will understand the meaning of quantization, quantum mechanical state and energy level. You will also understand that there is always a fundamental amount of kinetic energy, even for absolute zero temperature, $T = 0$ K, the so-called zero-point energy, which is responsible for the fact that matter is never at complete rest. Last, but not least, this experiment provides the basis for the chapters on vibrational spectroscopy.

quantum mechanical harmonic oscillator

4.2
Theoretical Background

4.2.1
Classical Versus Quantum Mechanics

In the previous experiment you have already learned how to determine the vibrational frequency of diatomic molecules using classical mechanics. It was only necessary to know the form of the interatomic potential close to the equilibrium position of the molecule and the masses of the two atoms. Classical mechanics, however, is lacking in two respects.

Computational Chemistry Workbook: Learning Through Examples
Thomas Heine, Jan-Ole Joswig, and Achim Gelessus
Copyright © 2009 WILEY-VCH Verlag GmbH & Co. KGaA, Weinheim
ISBN: 978-3-527-32442-2

First, it cannot predict the experimental fact that molecules cannot vibrate with any amplitude A as given in Equation 3.11; instead, the amplitude is quantized. It has been found that molecular vibrations can only absorb or emit energy in quanta $hf_0 = (h/2\pi)\omega_0 = \hbar\omega_0$, where f_0 is the frequency of the vibration and $h = 6.62608 \times 10^{-34}$ J s is the Planck constant. This way, molecular vibrations have the same energy as photons (light quanta) of the same frequency. Indeed, light is the standard means to activate or deactivate (by light absorption or emission) molecular vibrations, due to the coupling of the light's electromagnetic wave with the change of the molecule's dipole moment during the molecular vibration. As molecular vibrations have a characteristic frequency, the interaction of molecules with light provides the grounds for vibrational spectroscopy such as infrared (IR) and Raman spectroscopy. Both are key methods for the characterization of molecules.

For the second drawback of the classical treatment, it is harder to provide experimental evidence, at least for our subject of molecular vibrations. As we will see in this experiment, molecules will always have a minimum amount of kinetic energy. If a molecular vibration is completely deactivated, there is still energy amounting to $\frac{1}{2}hf_0 = \frac{1}{2}\hbar\omega_0$ in the system (see Figure 4.1). As this is the kinetic energy the system has **zero-point energy** at 0 K, it is called the zero-point energy (ZPE). As a consequence, the dissociation energy of the diatomic molecule described by the Morse potential (as in the previous chapter)

$$V(R) = D_e\left\{\left[1-\exp(-\alpha(R-R_e))\right]^2 - 1\right\} \qquad (4.1)$$

is not D_e, but in a quantum mechanical treatment it is $D_e - \frac{1}{2}\hbar\omega_0$. We may argue that the Planck constant is a tiny number, and in our macroscopic

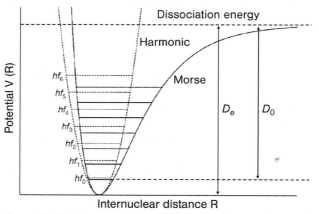

Figure 4.1 The Morse potential (solid line) and its harmonic approximation (dashed line) as in Figure 3.1. The vibrational levels for both potentials as well as the dissociation energy and the zero-point energy are indicated.

world it is impossible to observe the tiny difference of $\frac{1}{2}\hbar\omega_0$, but for a weakly bound molecule $\frac{1}{2}\hbar\omega_0$ is an appreciable value well within its potential scale, and the correction is indeed significant.

Physicists of the early twentieth century observed many more phenomena where unexpected things happened, in particular the photoelectric effect and the Franck–Hertz experiment. Both phenomena should be well understood by the student, and good explanations are available in most modern physics texts.

4.2.2
The Stationary Schrödinger Equation

These latter two phenomena can be explained by quantum mechanics, or quantum physics. It can be introduced in many different ways, and various texts are available for this purpose. The detailed exploration of this topic is strongly recommended to any modern scientist. For our experiment, we will start with the stationary Schrödinger equation for the special case of a single object – that means for the oscillating diatomic molecule. The stationary Schrödinger equation reads as

quantum mechanics

stationary Schrödinger equation

$$\hat{H}\psi = E\psi \tag{4.2}$$

and introduces some new quantities. Instead of giving a rigorous interpretation of Equation 4.2 and its variables, we will discuss them here using the example of the oscillations of a diatomic molecule. In the previous experiment, we learned that we can transform the equations of motion in such a way that the oscillations can be treated by a differential equation in one variable (the difference between the interatomic distance and the equilibrium distance, $\tilde{R} = R - R_e$) of a particle in a harmonic potential with the reduced mass μ (see Equation 3.6). We will now treat this problem with the stationary Schrödinger equation, written for our choice of variables as $\hat{H}\psi(\tilde{R}) = E\psi(\tilde{R})$.

The symbol $\hat{H} = \hat{T} + V$ denotes the Hamiltonian operator (often referred to simply as the Hamiltonian). The Hamiltonian is an energy operator, that is, upon operating it returns the energy of the system. It generally describes the system of electrons this is the electronic energy. In this case it is the total energy of the vibrating system, including the kinetic energy of the motion of the nuclei. The Hamiltonian contains two terms, the kinetic energy operator $\hat{T}$ and the potential V acting on the object, in this case the nuclei, using $\tilde{R}$ as principal variable (see Equation 3.10). The harmonic approximation of the Morse potential is

$$V = \frac{1}{2}\frac{\partial^2 V_{\mathrm{Morse}}(\tilde{R})}{\partial \tilde{R}^2}\bigg|_{\tilde{R}=0} \tilde{R}^2 = \frac{1}{2}K\tilde{R}^2 = \frac{1}{2}\mu\omega_0^2\tilde{R}^2 \tag{4.3}$$

It requires a longer introduction to motivate the analytical form of the kinetic energy operator $\hat{T}$, so we simply report the expression here and refer to the literature for a deeper discussion:

$$\hat{T} = -\frac{\hbar^2}{2\mu}\frac{d^2}{d\tilde{R}^2}$$

In Equation 4.2, E gives the energy of our object, the energy of the oscillation, composed of kinetic and potential contributions, just as in classical mechanics. The most remarkable new quantity is the wave-**wavefunction** function ψ, a function that characterizes the physical object, but without being a physical observable and hence without being accessible directly through experiments. The function is, however, directly connected to **probability density** the probability density:

$$p(\tilde{R}) = |\psi(\tilde{R})|^2 \tag{4.4}$$

In this experiment, we describe not a particle, but a molecular vibration, which is treated as a so-called quasi-particle with mass μ located in a linear coordinate system with axis $\tilde{R}$. In Equation 4.4, $p(\tilde{R})$ gives the probability of finding the particle at position $\tilde{R}$, that is, at the displacement from the equilibrium position.

The probability density $p(\tilde{R})$ can be used to calculate the probability that the molecule has a certain bond length. For example, to determine the probability that the molecule is within 1 pm of its equilibrium structure, we have to calculate the definite integral $\int_{-1\,\mathrm{pm}}^{1\,\mathrm{pm}} p(\tilde{R})\,d\tilde{R}$. We might find it useful to transform $\tilde{R}$ back to the bond length $R = \tilde{R} + R_e$. If we integrate the probability density over all interatomic distances, we should obtain

$$\int_{0}^{\infty} p(R)\,dR = 1 \tag{4.5a}$$

meaning literally that the molecule must have a bond length between zero and infinity at any time. Mathematically, we call Equation 4.5a **normalization condition** the normalization condition and we say that the probability density is a normalized function if it fulfills Equation 4.5a. Indeed, we will always request the normalization condition to hold when we solve the Schrödinger equation. Since we will use the displacement from the equilibrium bond length $\tilde{R}$ in the following, we convert Equation 4.5a to

$$\int_{-\infty}^{\infty} p(\tilde{R})\,d\tilde{R} = 1 \tag{4.5b}$$

Note that Equation 4.5b formally includes also negative bond lengths, but the integral over the entire X axis is necessary and is an artifact of the harmonic approximation. Remember that the results are only valid in the vicinity of $\tilde{R} = 0$.

4.2.3
Solutions to the Schrödinger Equation

Let us now solve the Schrödinger equation for our problem and discuss the quantities of Equation 4.2, the subsequent paragraphs and their results. In summary, we have to solve the following partial differential equation:

$$-\underbrace{\frac{\hbar^2}{2\mu}\frac{d^2}{d\tilde{R}^2}}_{\hat{T}}\psi(\tilde{R}) + \underbrace{\left(\frac{1}{2}\mu\omega_0^2\tilde{R}^2}_{V} - E\right)\psi(\tilde{R}) = 0 \qquad (4.6)$$

The solution of Equation 4.6 is not straightforward, but we are only interested in its result at the moment. The analytic solution can be followed in various texts on quantum physics, but it can also be calculated using a modern math processor. In contrast to classical mechanics, where the solution to the problem is one characteristic angular frequency for the vibration ω_0, but oscillations with any amplitude A are possible, the Schrödinger equation yields an infinite number of discrete solutions for the wavefunction $\psi_n(\tilde{R})$. The quantum number n can take integer values $n = 0, 1, 2, \ldots$. The solutions are related to the Hermite polynomials, a series of mathematical functions with the form

quantum number

Hermite polynomials

$$H_n(x) = (-1)^n e^{x^2}\frac{\partial^n e^{-x^2}}{\partial x^n} \qquad (4.7)$$

which are defined for $n = 0, 1, 2, \ldots$. The rather longish analytical form of the solutions of Equation 4.6 is

$$\psi_n(\tilde{R}) = N_n \exp\left(-\frac{\mu\omega_0}{2\hbar}\tilde{R}^2\right) H_n\left(\sqrt{\frac{\mu\omega_0}{\hbar}}\tilde{R}\right) \qquad (4.8)$$

The normalization condition, Equation 4.5b, is satisfied if we choose the normalization factor N_n to be

$$N_n = \left(\frac{\mu\omega_0}{\pi\hbar}\right)^{1/4}\frac{1}{\sqrt{2^n n!}} \qquad (4.9)$$

Each solution $\psi_n(\tilde{R})$ to Equation 4.6 yields a different value for the energy, which we consequently label according to the wavefunction with

$$E_n = (n + \tfrac{1}{2})\hbar\omega_0 = (n + \tfrac{1}{2})hf_0, \quad n = 0, 1, 2, \ldots \qquad (4.10)$$

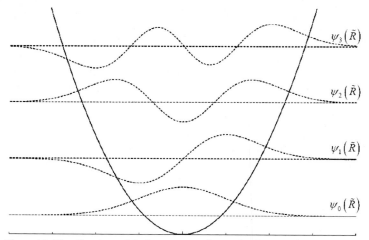

Figure 4.2 Wavefunctions of the harmonic oscillator. The abscissas for drawing the individual wavefunctions $\psi_n(\tilde{R})$ are shifted to their respective energy levels (each by $\frac{1}{2}\hbar\omega_0$).

quantization   Equation 4.10 gives the quantization of energy. It states that only discrete energy values are allowed for the system. The solutions of the stationary Schrödinger equation, Equation 4.6, are visualized in Figure 4.2. The results in Equations 4.8 to 4.10 are in many respects spectacular. They stop us from applying various concepts that we know from daily life and classical physics. Many of the remarkable implications of quantum physics are important for chemistry. We will highlight some examples later.

Let us start with the energy of the vibration. In the case of a classical vibration, the energy is given as the sum of potential and kinetic energy. It is directly related to the amplitude of the vibration, at the reversal points where $\tilde{R}$ is at its extremum $\tilde{R}_{max}$, the kinetic energy is zero, and we can calculate the total energy as $E_{tot} = \frac{1}{2}\mu\omega_0^2 \tilde{R}_{max}^2$. Obviously, for energetic reasons, the molecule cannot have an interatomic distance outside the interval $R_e - \tilde{R}_{max} \leq R \leq R_e + \tilde{R}_{max}$. In this classical picture, any value is possible for the total energy, as the amplitude of the oscillation can be chosen freely. This is entirely different in quantum physics. Equation 4.10 states that only selected energy levels E_n are allowed for **eigenenergy**   the molecular vibrations. We call each of these energies an eigenenergy.[1] The eigenenergies differ by quanta of $\hbar\omega_0 = hf_0$, and these differences are observed experimentally, since molecules can absorb

---

1) The word "eigen" is German and translates as "own". A possible translation for eigenenergy is the system's inherent or intrinsic energy.

and emit photons of characteristic energies only. This is the fundament of all vibrational spectroscopy (see also Chapters 11 and 12). An important fact is the presence of a finite energy for the lowest eigen-energy ($n = 0$): the system still carries a vibrational energy of $\frac{1}{2}\hbar\omega_0$. This is not understandable within classical physics, where the lowest energy state is simply a system that is at rest (R is at R_e for all times). Eigenenergy and ware function with the same quantum number n are called a quantum state or simply "state". The state with lowest allowed **quantum state** quantum number n is called "ground state". **ground state**

The list of surprises is not restricted to the vibrational energy levels. The possible values for the interatomic distance are also interesting. Let us calculate the probability function for the first three vibrational levels. First, we calculate the Hermite polynomials according to Equation 4.7 and obtain the polynomials for the first three states:

$$
\begin{aligned}
H_0(x) &= 1 \\
H_1(x) &= 2x \\
H_2(x) &= 4x^2 - 2
\end{aligned}
\tag{4.11}
$$

We insert Equations 4.11 into Equation 4.8 and obtain for the wave-function of the first three states:

$$
\psi_0(\tilde{R}) = \left(\frac{\mu\omega_0}{\pi\hbar}\right)^{1/4} e^{-\frac{\mu\omega_0}{2\hbar}\tilde{R}^2}
$$

$$
\psi_1(\tilde{R}) = \left[\frac{4}{\pi}\left(\frac{\mu\omega_0}{\hbar}\right)^3\right]^{1/4} e^{-\frac{\mu\omega_0}{2\hbar}\tilde{R}^2}\tilde{R}
\tag{4.12}
$$

$$
\psi_2(\tilde{R}) = \frac{1}{2\sqrt{2}}\left(\frac{\mu\omega_0}{\pi\hbar}\right)^{1/4} e^{-\frac{\mu\omega_0}{2\hbar}\tilde{R}^2}\left(4\frac{\mu\omega_0}{\hbar}\tilde{R}^2 - 2\right)
$$

As they are characteristic for the problem, we speak of eigenstates or eigenfunctions of the problem. According to Equation 4.4 the **eigenfunction** probability densities (see Figure 4.3) of the first three states are consequently

$$
p_0(\tilde{R}) = \sqrt{\frac{\mu\omega_0}{\pi\hbar}}\ e^{-\frac{\mu\omega_0}{2\hbar}\tilde{R}^2}
$$

$$
p_1(\tilde{R}) = \sqrt{\frac{4}{\pi}\left(\frac{\mu\omega_0}{\hbar}\right)^3}\ e^{-\frac{\mu\omega_0}{2\hbar}\tilde{R}^2}\tilde{R}
\tag{4.13}
$$

$$
p_2(\tilde{R}) = \frac{1}{8}\sqrt{\frac{\mu\omega_0}{\pi\hbar}}\ e^{-\frac{\mu\omega_0}{2\hbar}\tilde{R}^2}\left(4\frac{\mu\omega_0}{\hbar}\tilde{R}^2 - 2\right)^2
$$

If we inspect Equations 4.13 carefully we realize that the probability density is enveloped by a Gaussian function, which decays exponentially with $\tilde{R}^2$. This means that the probability is never exactly zero, so

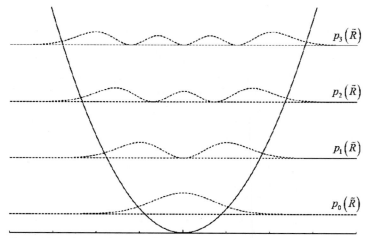

Figure 4.3 Probability densities of the harmonic oscillator. Conventions are as in Figure 4.2.

that there is always a finite probability to have the molecule with an interatomic distance significantly different from the equilibrium distance. This is most peculiar, as it means that there is a finite probability for the system to have an interatomic distance corresponding to a potential energy that exceeds the total energy of the system! This property of quantum physics has been proven experimentally: it is

tunneling the basis of the tunnel effect, or simply tunneling, where a quantum particle can cross a potential barrier that exceeds its energy. We may argue that the result would imply unphysical results, as it would suggest the possibility of negative interatomic distances, but this result is an artifact of the harmonic approximation. The true potential goes to infinity as $R \rightarrow 0$, and hence the probability function is zero for the true potential.

In vibrational spectroscopy, molecules interact with photons. They can absorb several photons, which allow the vibration to be excited from level $0 \rightarrow 1 \rightarrow 2 \rightarrow \dots$. We therefore speak of the nth excitation of a vibration if the vibration is in state n.

Let us remember that the quality of the results depends crucially on the quality of the harmonic approximation. We have already observed an artifact of this approximation: it gives finite probabilities for non-existing bond lengths smaller than zero. Especially, for higher vibrational levels, the harmonic approximation will give less accurate results. It yields equally spaced eigenstates and symmetric probability densities. For the true potential the distance between energy levels gets smaller for higher n, and the probability density gets confined towards smaller, but stretched towards larger, interatomic distances (see Figure 4.1).

4.3
Demonstration

Again, we investigate the vibrational frequency of the carbon monoxide (CO) molecule. The potential for the harmonic approximation is determined as in the Demonstration of Chapter 3, and we obtain the values $K = 1.147 \times 10^{27}\,\mathrm{kg\,mol^{-1}s^{-2}}$, $\mu = 6.86\,\mathrm{g\,mol^{-1}}$ and $\omega_0 = \sqrt{K/\mu} = 4.09 \times 10^{14}\,\mathrm{s^{-1}}$. Within the harmonic approximation, the zero-point energy is $E_{ZPE} = \frac{1}{2}\hbar\omega_0 = 13\,\mathrm{kJ\,mol^{-1}}$. The vibration has an energy difference between two states of $26\,\mathrm{kJ\,mol^{-1}}$ and the molecule can, thus, absorb or emit a photon of the same energy, $E_{photon} = hf_{photon} = \Delta E_{vibration}$. From the photon's frequency we can now calculate its wavelength:

$$\lambda_{photon} = \frac{hc}{E_{photon}} = \frac{2\pi c}{\omega_{vibration}} = 4608\,\mathrm{nm}$$

CO has a bond length of $1.13\,\text{Å} = 1.13 \times 10^{-10}\,\mathrm{m} = 113\,\mathrm{pm}$. The probability p of finding the bond length within a variation of $\pm\Delta = \pm 10^{-12}\,\mathrm{m}$, that is, between 112 and 114 pm, in the molecule's vibrational ground state, that is, in the lowest energy vibration, is given (cf. Equation 4.13) by

$$p = \sqrt{\frac{\mu\omega_0}{\pi\hbar}} \int_{-\Delta}^{+\Delta} e^{-\frac{\mu\omega_0}{\hbar}\tilde{R}^2}\,d\tilde{R}$$

$$= 2\sqrt{\frac{\mu\omega_0}{\pi\hbar}} \int_{0}^{\Delta} e^{-\frac{\mu\omega_0}{\hbar}\tilde{R}^2}\,d\tilde{R}$$

Here, we already make use of the fact that the vibration is symmetric with respect to $\tilde{R} = 0$. For clarity, we substitute $X = \sqrt{(\mu\omega_0/\hbar)}\tilde{R}$, which also gives

$$\frac{dX}{d\tilde{R}} = \sqrt{\frac{\mu\omega_0}{\hbar}} \quad \text{and} \quad dX = \sqrt{\frac{\mu\omega_0}{\hbar}}d\tilde{R}$$

Now we end up with

$$p = \frac{2}{\sqrt{\pi}} \int_{0}^{\Delta} e^{-\frac{\mu\omega_0}{\hbar}\tilde{R}^2}\sqrt{\frac{\mu\omega_0}{\hbar}}\,d\tilde{R}$$

$$= \frac{2}{\sqrt{\pi}} \int_{0}^{\sqrt{(\mu\omega_0/\hbar)}\Delta} e^{-X^2}\,dX$$

$$= \mathrm{erf}\left(\sqrt{\frac{\mu\omega_0}{\hbar}}\Delta\right) = \mathrm{erf}(0.228) = 0.253$$

that is 25.3%.

Next, we are going to calculate the classical amplitude of the ground-state vibration. We use the vibrational energy of the ground state and combine it with the classical expression for the total energy of the vibration (the energy at the reversal point $\tilde{R}_{rev} = \tilde{R}_{max}$):

$$\frac{1}{2}\hbar\omega_0 = \frac{1}{2}k\tilde{R}^2_{rev} \quad \Rightarrow \quad \tilde{R}_{rev} = \sqrt{\hbar\omega_0/K} = 4.8 \text{ pm}$$

To calculate the probability of finding the particle outside its classical radius within the harmonic approximation, we have to integrate over $\tilde{R}$ with $\tilde{R}_{rev} \leq \tilde{R} \leq \infty$:

$$p = \sqrt{\frac{\mu\omega_0}{\pi\hbar}} \int\limits_{\tilde{R}_{rev}}^{\infty} e^{-\frac{\mu\omega_0}{\hbar}\tilde{R}^2} \, d\tilde{R}$$

$$= \frac{1}{2}\text{erf}\left(\sqrt{\frac{\mu\omega_0}{\hbar}}\tilde{R}_{rev}\right) = 0.078$$

This means that there is a 15.6% probability of finding the particle outside its classical limits for the ground state (7.8% on each side).

4.4
Problems

1. Zero-point energy

Calculate the zero-point energies for hydrogen (H_2), fluorine (F_2), chlorine (Cl_2), bromine (Br_2), and iodine (I_2). The Morse potential parameters are given in Table 3.1. Calculate the contribution of the zero-point energy to the binding energy in percent.

2. Probabilities

For each of the molecules of problem 1, calculate the probability of the molecule having an interatomic distance (a) within 1 pm of the equilibrium bond length, and (b) between its outer classical reversal point and the dissociation limit if the vibration is not excited ($n = 0$).

3. Transformation of light to molecular vibration

Calculate the wavelength of light that is absorbed in the process to activate a vibrational level in the molecules of problem 1.

4. Orthonormality

Show that the wavefunctions for $n = 0, 1, 2$ for the molecular vibrations of H_2 are orthogonal to each other and normalized.

4.5
Technical Details

For this task, a spreadsheet (Open Office Calc as on the CD, Microsoft Excel) will reduce the calculus-related work significantly.

4.6
Review and Summary

We have treated the vibrations of diatomic molecules, using quantum mechanics. These vibrations are so-called quasi-particles. For that purpose we introduced the stationary Schrödinger equation. Further reading on the fundamentals of quantum physics is encouraged.

In contrast to the classical treatment, the Schrödinger equation includes, besides the interatomic potential, the kinetic energy of the system, and leads to partial differential equations that have to be solved to calculate the wavefunction and the energy of the system. As for the classical treatment, we approximated the potential around the equilibrium position with a parabola (harmonic approximation), and treated the system in one dimension along its bond. Here, we could use the same mathematical and physical apparatus as in Chapter 3. The solution of the Schrödinger equation is the quantum mechanical harmonic oscillator, which is a good model of the vibrations of diatomic molecules. We obtained a series of solutions for the wavefunction, and consequently a series of probability densities of the vibrations and vibrational energies. Only a discrete set of energies is allowed for the system, which implies that only finite energy differences between different vibrational states are possible.

We observed fundamental differences of quantum mechanical solutions to classical physics: the probability densities of the vibrational states allow values that are outside the classical amplitudes, and even in the ground state the system carries kinetic energy, the zero-point energy. We quantified these effects for a series of example molecules of different atomic masses and interaction potentials.

References

1 Atkins, P.W. and de Paula, J. (2006) *Physical Chemistry*, 4th edn, Oxford University Press, Chapter 13.

5
Atomic Orbitals

5.1
Aim

In this computer experiment, you become familiar with the solution of the electronic Schrödinger equation for free atoms. To avoid complex math and physics, you are introduced by an empirical approach to atomic orbitals, probability densities of electrons, energy levels, quantum numbers and so on. Using the tools of modern quantum chemistry, you calculate the energy levels of free atoms and atomic orbitals, and you employ computer graphics for visualizing them. Using the graphical interface, you develop your understanding of energy and length scales in atoms and molecules. Most of all, you compare the shape of different orbital types, the core, valence, and unoccupied orbitals, a necessary precondition to understand chemical bonding. As many readers of this book are expected to have limited knowledge of quantum physics, the fundamentals are briefly reviewed and, more importantly, the essentials for the chemist are summarized. This experiment is accompanied by some calculus examples to practice your understanding.

5.2
Theoretical Background

5.2.1
The Schrödinger Equation of the Atom

One of the problems that physics majors learn during their quantum mechanics course is to solve the Schrödinger equation for the hydrogen atom. Indeed, this is one of the few cases where a real system can be treated analytically within reasonable approximations. For most chemists it is not essential to be able to follow each step in this demanding mathematical procedure, but today it is important to know that we can

electronic Schrödinger equation

Computational Chemistry Workbook: Learning Through Examples
Thomas Heine, Jan-Ole Joswig, and Achim Gelessus
Copyright © 2009 WILEY-VCH Verlag GmbH & Co. KGaA, Weinheim
ISBN: 978-3-527-32442-2

solve the Schrödinger equation with great precision numerically, using various adequate approximations on a modern computer together with modern nummerical tools. At this stage, it is important to develop a **atomic orbital** feeling about the shape and the range of atomic orbitals, and of the **orthogonality** figurative meaning of orthogonality and normalization, as this is the **normalization** precondition for various bonding models: molecular orbital theory and the linear combination of atomic orbitals (LCAO) scheme, both explained in Chapter 7.

If you want to follow the solution of the Schrödinger equation step by step, you are referred to one of the many texts dealing with this topic in detail, for example [1, 2]. You should remember, however, that as soon as we have to deal with more than one electron in the system, the nice analytical approach fails and we are forced to solve the Schrödinger equation numerically. In this computer experiment, we employ density-functional theory [3], but other methods would work just as well.

The stationary Schrödinger equation yields an infinite number of solutions $\psi_{nlm}(r)$, called by various terms like orbitals, states or one-**quantum number** particle wavefunctions. The solutions are labeled by the integer quantum numbers n,l,m, and each state is associated with a state energy ε_{nlm}. It is now important to pick the important states and to discuss their properties. To do that, we need to accept a few facts that you will possibly learn in a theoretical chemistry or quantum physics course.

1. The wavefunctions, that is, the solution of the Schrödinger equation of one electron in the vicinity of a point charge and some mean field the other electrons, are called *orbitals* $\psi_{nlm}(r)$. Here r is the position vector of the electron. Mathematically, they are scalar functions in three dimensions $r = (x, y, z)$. An infinite number of solutions is found. As the atom is in three-dimensional space, they are naturally labeled by three quantum numbers n,l,m, which are known as the principal (main), the angular momentum, and the magnetic quantum numbers, respectively. Each solution is assigned an orbital **orbital energy** energy ε_{nlm}.

2. Only certain integer values are allowed for each quantum number. **principal quantum number** The principal quantum number n is given by the series of natural **angular momentum** numbers $1,2,3,\ldots$. The range of the angular momentum quantum **number** number l depends on the principal quantum number n and can take any integer value between 0 and $n-1$; that is, for $n=1$ the only allowed value is $l=0$, for $n=2$ there are two allowed values $l=0$ and $l=1$, and so forth. Alternatively, angular momentum quantum numbers are denoted by letters ($s,p,d,f,\ldots$ corresponding to l **magnetic number** $0,1,2,3,\ldots$). The range of the magnetic quantum number m depends on l and can take any integer value between $-l$ and $+l$, that is, $m = -l, \ldots, 0, \ldots, +l$.

3. The orbitals are occupied by electrons, starting from the orbital of lowest energy (Aufbau[1] principle). The low-energy solutions are those with the smallest quantum numbers n,l. For the H atom, the orbital energy depends only on the principal quantum number and is

$$\varepsilon_n = -\frac{1}{2n^2}$$

(we will use Hartree as the energy unit in this experiment). In atomic units the energy is given in Hartree (1 Hartree = 2 Rydberg = 27.211 eV), the length is Bohr radii ($1a_0 = 5.29177 \cdot 10^{-11}$m), and the mass in electron masses ($1m_e = 9.10938 \cdot 10^{-31}$ kg). For atoms with more than one electron, the relation is far more complex and cannot be given in analytic terms. However, the low-energy orbitals are those with small quantum numbers. There are well-known exceptions for larger atoms: for example, ε_{4s} is often lower than ε_{3d}.

4. We calculate solutions for electrons. Electrons are elementary particles and belong to the family of fermions, which can be characterized by only two different intrinsic states, the so-called particle spin. In this experiment we discuss only solutions of noble gas atoms. Here, we always have the same number of spin-up and spin-down electrons, which occupy states of the same energy. This simplifies the problem such that each state n,l,m either is occupied by two electrons or is empty, which is reflected in the occupation number,

$$n_{nlm} = \begin{cases} 2 \text{ for occupied states} \\ 0 \text{ for unoccupied states} \end{cases}$$

Atoms and molecules that have this property are diamagnetic and called closed-shell systems.

5.2.2
Atomic Orbitals

The main purpose of this experiment is that you familiarize yourself with the meaning of orbitals. As we are discussing stationary problems (those where the system is not changing in time), the orbitals describe, now the space is occupied by the electrons. The orbitals exhibit several important properties, as follows.

orbital

1. The square of the modulus of an orbital ψ_{nlm} is the probability density $p_{nlm}(x,y,z) = |\psi_{nlm}|^2(x,y,z)$. Note that the wavefunction is complex in general, but for all the cases we discuss in this book it is real. This quantity is useful to determine the probability of finding an electron in a certain part of space, for example in a sphere with

---

[1] "Aufbau" is German and literally translates to "build up".

radius 1 Å around the nucleus. Obviously, if we integrate the orbital probability density over the entire space, we obtain unity, which means that the electron is present at least somewhere in the universe. In mathematical terms, the probability of finding the electron of the occupied state ψ_{nlm} in volume ΔV is

$$p_{nlm}(\Delta V) = \iiint_{\Delta V} \psi_{nlm}{}^2(x, y, z)\, dx\, dy\, dz$$

Integration over all space is given by

$$\int_{-\infty}^{\infty} \int_{-\infty}^{\infty} \int_{-\infty}^{\infty} \psi_{nlm}{}^2(x, y, z)\, dx\, dy\, dz = 1$$

This property is known as *normalization*.

2. The sum of the probability densities of all the occupied orbitals sums up to the total electron density:

$$\rho(r) = \rho(x, y, z) = \sum_{n,l,m} n_{nlm} |\psi_{nlm}|^2$$

This function gives the distribution of electrons in the atom. Integrating it over all space gives the number of electrons in the system:

$$N = \sum_{n,l,m} \int_{-\infty}^{\infty} \int_{-\infty}^{\infty} \int_{-\infty}^{\infty} n_{nlm} |\psi_{nlm}|^2 (x, y, z)\, dx\, dy\, dz$$

$$= \int_{-\infty}^{\infty} \int_{-\infty}^{\infty} \int_{-\infty}^{\infty} n_{nlm} \rho(x, y, z)\, dx\, dy\, dz$$

3. As solutions to the Schrödinger equation, the orbitals are *orthogonal* to each other. This orthogonality is the quantum physical correspondence to the classical law that there can be no object at the position of another one. Looking at the probability density of any atomic state, one realizes that the wavefunction decays away from the nucleus with exponential asymptotics. That means that, everywhere in space, there is a finite probability of finding the electron of the state, and at this position we must assure that no two electrons of two different states must be present at the very same time. Mathematically, this is ensured if the integral of the product of two different states (at least one index n, l, m differs from n', l', m') over space vanishes, that is

$$\int_{-\infty}^{\infty} \int_{-\infty}^{\infty} \int_{-\infty}^{\infty} \psi_{nlm}(x, y, z)\, \psi_{n'l'm'}^{*}(x, y, z)\, dx\, dy\, dz = 0$$

We can combine this equation with the normalization condition and obtain the orthonormalization condition for the orbitals:

orthonormalization

$$\int\limits_{-\infty}^{\infty} \int\limits_{-\infty}^{\infty} \int\limits_{-\infty}^{\infty} \psi_{nlm}(x,y,z)\psi_{n'l'm'}^{*}(x,y,z)\mathrm{d}x\,\mathrm{d}y\,\mathrm{d}z = \delta_{nn'}\delta_{ll'}\delta_{mm'}$$

where δ_{ij} denotes the Kronecker symbol defined as

Kronecker symbol

$$\delta_{ij} = \begin{cases} 1 & \text{for } i = j \\ 0 & \text{for } i \neq j \end{cases}$$

The spatial properties of the atomic orbitals will be evident when you carry out the Demonstration and solve the problems.

5.3
Demonstration

We will start with the argon atom. We learn from the periodic table that argon has 18 electrons and the electron configuration $1s^2 2s^2 2p^6 3s^2 3p^6$. We solve the approximate Schrödinger equation for this atom. If you want to use the software delivered with this book, consult the Technical Details in Section 5.4 for technical advice.

After solving the Schrödinger equation, you obtain orbitals and orbital energies. Let us concentrate first on the orbital energies. The *1s* orbital has an energy of approximately −113.9 Hartree. Realize the significance of this quantity. It means that the binding energy of the 1s electrons in argon is more than 200 times higher than the binding energy of an electron in the hydrogen atom, which is exactly 0.5 Hartree. It is evident that those electrons are strongly trapped in their states and will not contribute to any type of chemical bond. The same conclusions are true for the *2s* orbital (at around −10.8 Hartree) and the three *2p* orbitals (at around −8.4 Hartree). This is also the reason why the electrons of these orbitals are called core electrons. Core electrons are strongly bound and confined to the nucleus and have a negligible contribution to any type of chemical bond.

The next orbitals are the valence orbitals with orbital energies comparable with that of the hydrogen atom (*3s* at −0.9 and the three *3p* at about −0.4 Hartree). These numbers confirm that the valence electrons are the most important ones for chemical reactions – the core electrons are so strongly bound that their activation is not the business of synthetic chemists.

core electrons

The next step is to evaluate the orbital shapes of the atomic orbitals. Let us start with the lowest energy orbital, $\psi_{100}(\mathbf{r})$ (Figure 5.1a). This orbital has a spherical shape, which means that points of equal values of the wavefunction are located on spheres (we are looking at a so-called

orbital shape
molecular orbital

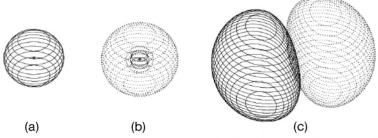

(a)   (b)   (c)

Figure 5.1 The (a) *1s*, (b) *2s*, and (c) *2p$_x$* orbitals of the argon atom. The two signs of the wavefunction are indicated by solid and dotted lines.

isosurface), and the wavefunction only approaches zero for very large distances from the nucleus. The next orbital, $\psi_{200}(r)$ (Figure 5.1b), is already more complex, even though it still obeys spherical symmetry. It

node   has a node, which means a sphere with radius $|r|$ where the values of $\psi_{200}(r)$ are zero and the wavefunction changes sign with respect to the distance from the nucleus. You can see the node, depending on your visualization software, by changing the value for the isosurfaces. The orbitals $\psi_{21-1}(r)$, $\psi_{210}(r)$, and $\psi_{211}(r)$ have a clear directional dependence (Figure 5.1c). All of these three orbitals have nodes, this time for planes perpendicular to the direction of the orbital "clubs" crossing the nucleus. The orbitals are antisymmetric functions in one space coordinate, and for convenience they are usually aligned in the *x*, *y*, and *z* directions.

Now we proceed to the third principal quantum number. The $\psi_{300}(r)$ orbital (Figure 5.2a) is again spherically symmetric, but now it has two radial nodes and changes sign twice. Note that the orbital has a much greater spatial extension than the lowering orbitals below, and that there is no node in the bonding area. We make a similar observation for the *3p* shell ($\psi_{31-1}(r)$, $\psi_{310}(r)$, and $\psi_{311}(r)$, Figure 5.2b). There are now not only nodal planes through the nucleus, but also club-shaped nodes inside each club (Figure 5.2), the spatial extension is much greater than for the *2p* shell, and the orbitals are antisymmetric in one direction of space.

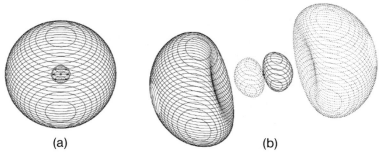

(a)   (b)

Figure 5.2 The (a) *3s* and (b) *3p$_x$* orbitals of the argon atom. The two signs of the wavefunction are indicated by solid and dotted lines.

With these observations, we understand that chemistry in different shells remains similar, even though the spatial dimensions of the bonds vary due to the different atomic sizes. In a convenient picture, we understand that in molecules the electrons of the valence orbitals of different atoms interact with each other. The interaction is expressed by the overlap of the valence orbitals of neighboring atoms. As there are no nodes in the atomic orbitals in the space where chemical bonds are formed, the concept of bonding can be kept simple. If we consider only the valence electrons with their (l,m) characteristics, we have an approximate description of the chemistry in a molecule, that is still qualitatively, often even quantitatively, correct.

5.4
Problems

1. The electron configuration of atoms

Draft the electron configurations of helium, neon, argon and krypton in occupation diagrams as you would have done in your basic chemistry course.

2. The energy levels of an atom

Solve the stationary Schrödinger equation of the atoms of problem 1 as approximated in your computational chemistry software. Now draw the same diagrams as in problem 1 using the orbital energies obtained by the calculation. Note the scale of the binding energies of electrons in atoms and compare it with the energy scale of the valence electrons. Note also the energy splitting of the principal quantum number and the angular momentum quantum number.

3. The shape of the valence orbitals

Visualize the valence orbitals of the four noble gas atoms. Compare the spatial extension of the valence orbitals of the same angular momentum and magnetic quantum numbers.

4. Molecular orbitals in three dimensions

Spend some time visualizing the orbitals that cannot be drawn too easily in two dimensions on paper, in particular the d orbitals.

5. Orbital form and bonding types

Which bonding types can be realized by s, p and d orbitals? Rationalize your prediction.

6. Orthonormality of the atomic orbitals

Solve the following calculus problems, which will help you to understand the concept of orthogonality and normalization. The orbital wavefunctions of the *1s*, *2s* and *2p$_x$* states of the neon atom can be approximated by the following analytical expressions:

$$\psi_{1s} = N_{1s}e^{-9.18297r}$$

$$\psi_{2s} = N_{2s}(-5.81098e^{-7.18832r} + 1.85787e^{-1.96677r})$$

$$\psi_{2p_x} = N_{2p}xe^{-2.71225r}$$

For each of these orbitals, calculate the nodes. Calculate the radius of maximum probability density of the 1s orbital (more advanced: of all three orbitals). Investigate if the three orbitals are orthogonal to each other. Finally, determine the normalization coefficients of the two s orbitals N_{1s} and N_{2s} (more advanced: also N_{2p}). You may find the following integral useful:

$$\int_0^\infty r^n e^{-ar} dr = \frac{n!}{a^{n+1}}$$

5.5
Technical Details

For problems 2–5 you can use the software on the CD delivered with this book. First, you need to solve the stationary Schrödinger equation for the atoms. For the deMon program, an example input for the neon atom (which might be named ne.inp) looks as follows:

```
TITLE NE ATOM
VXCTYPE AUXIS PBE
BASIS (DZVP-GGA)
AUXIS (GEN-A2*)
VISUALISATION MOLDEN
GEOMETRY CARTESIAN
Ne 0.0 0.0 0.0
END
```

The atomic orbitals can be visualized using Molden, also present on the CD. For this purpose, interface the .mol file that has been generated by deMon with Molden, for example, by typing molden ne.mol. You can draw the atomic orbitals using the **Dens. Mode** button.

5.6
Review and Summary

The quantum character of electrons in atoms has been introduced. Students learn that atomic orbitals are a solutions of the stationary

Schrödinger equation of electrons in an external potential of a single nucleus. The shell structure of the electrons became visible both mathematically as well as on the energy scale, and the assignment of quantum numbers to the orbitals has been motivated.

Atomic orbitals have not only been introduced figuratively by plotting them, but also mathematically. They are orthogonal to each other, which means that the integral of the wavefunction products over all space is zero corresponding to the classical picture that two objects can not be at the same point at the same time. Moreover, atomic orbitals are normalized, so that the integral of the squared wavefunction over all space is 1. By direct comparison, you have seen that orbitals of the same quantum number, but belonging to different elements, may have extremely different extensions: Core orbitals are bound significantly tighter than valence orbitals. Therefore, chemical bonding is determined in most cases by the valence electrons only. Finally, you got a first introduction into computational software which we will use much more intensively in the forthcoming chapters.

References

1 Born, M. (1989) *Atomic Physics*, 8th rev. edn, Dover Publications.
2 Atkins, P.W. and de Paula, J. (2006) *Physical Chemistry*, 4th edn, Oxford University Press.
3 Koch, W. and Holthausen, M.C. (2001) *A Chemist's Guide to Density-Functional Theory*, 2nd edn, Wiley-VCH Verlag.

6
Ionization Potentials and Electron Affinities of Atoms

6.1
Aim

We learn to calculate ionization potentials and electron affinities of atoms of the second period. By applying the Slater rules we will see that both quantities can be approximated by evaluation of the electronic shielding of the nucleus. Using quantum chemistry we will see that ionization potentials can be estimated in very good agreement with experiment, and even electron affinities can be predicted with good accuracy for stable anions if the numerical details are chosen carefully. Comparison of the calculated ionization potentials and electron affinities of the elements of the second period reveals the chemical character of these elements.

6.2
Theoretical Background

6.2.1
Ionization Potential and Electron Affinity

Two important quantities for the reactivity of atoms and molecules are ionization potential (IP) and electron affinity (EA). Chemically inert substances are hard to ionize and they accept electrons only with a loss of energy. We can define both quantities in terms of chemical reactions.

The ionization potential is the lowest energy that is necessary to **ionization potential** remove one electron from substance A:

$$A \rightarrow A^+ + e^-$$

After this reaction, ion A^+ and electron e^- are separated by an infinite distance and do not interact. The total energy of the free electron e^- is zero, so the ionization potential can be calculated as the difference

Computational Chemistry Workbook: Learning Through Examples
Thomas Heine, Jan-Ole Joswig, and Achim Gelessus
Copyright © 2009 WILEY-VCH Verlag GmbH & Co. KGaA, Weinheim
ISBN: 978-3-527-32442-2

between the total energies[1] of the neutral atom and the ion:

$$IP = E_{tot}(A^+) - E_{tot}(A) \tag{6.1}$$

HOMO   A simple picture illustrates this reaction. An electron is transferred from the highest occupied molecular orbital (HOMO) of the neutral particle A to vacuum. Assuming that the other electrons of A are not further perturbed, the energy that is necessary to ionize A is the binding energy of the most weakly bound electron of A, that is, the energy of the HOMO:

$$IP \approx -\varepsilon_{HOMO} \tag{6.2}$$

Koopmans' approximation   This approximation is also known as Koopmans' approximation or Koopmans' theorem [1]. It is interesting to note that Tjalling Charles Koopmans (1910–1985) worked on Hartree–Fock theory at the Theoretical Physics departments of Utrecht University, The Netherlands, before he changed to economics. For his contributions to the latter field, in 1975 he was awarded the Sveriges Riksbank Prize in Economic Sciences in Memory of Alfred Nobel (the "Nobel Prize in Economics").

first isolation potential
second isolation potential   The energy to remove one electron from a neutral species is more precisely called the first ionization potential. Higher ionization potentials denote the further ionization of ions, for example, the second ionization potential is given by

$$IP_2 = E_{tot}(A^{2+}) - E_{tot}(A^+) \tag{6.3}$$

electron affinity   The electron affinity is related to the opposite process: it is the binding energy of an extra electron to a neutral system. The chemical reaction is

$$A + e^- \rightarrow A^-$$

Again, the free electron has zero energy. The sign convention of the electron affinity is opposite to the usual one in thermodynamics, so the electron affinity is calculated as the difference between the energies of the neutral system and its anion:

$$EA = E_{tot}(A) - E_{tot}(A^-) \tag{6.4}$$

For stable atomic anions, the results are positive. Indeed, all electron affinities of atoms reported by modern experiments are positive.

---

[1] "Total energy" is a term that is only partially correct. It denotes the electronic energy and the nucleus–nucleus repulsion, so it is the total energy of a frozen molecule. In the chapters where the kinetic energy of the nuclei is calculated, we will refer to this number more correctly as the "potential energy". But as the term "total energy" is also used in most computational chemistry software, we continue to use it here. Note that within this convention the total energy of atoms equals their electronic energy.

However, in the older literature some negative results can be found, which should be discarded. For some elements, for example, beryllium (Be) or magnesium (Mg), the anionic form is instable, and no electron affinity can be measured. Calculation may predict negative numbers for those cases, which should be interpreted such that the anionic form is unstable and cannot be formed according to this level of theory.

6.2.2
Slater Rules: Wavefunction and Binding Energies of Electrons in Atoms and Ions

In 1930, John C. Slater (1900–1976) suggested some simple, but effective, rules for the calculation of electron energies in atoms, the so-called Slater rules [2]. With these rules, ionization potentials, and to some extent electron affinities, can be estimated. We will judge these estimates later in this chapter.   **Slater rules**

Slater started by giving an approximate analytic expression for atomic orbitals. To avoid the treatment of the explicit N-electron problem, he instead described single electrons in an approximate central field, created by the nucleus and the remaining $N - 1$ electrons. As the radial nodes are usually close to the nucleus, they do not significantly affect properties such as orbital energies and are therefore neglected. The radial part R_{nl} of the atomic orbital as a function of the distance from the nucleus (in Bohr radii $\rho = r/a_0$) is given by

$$R_{nl}(\rho) = A\rho^{n^*-1}e^{-\frac{Z-s}{n^*}} \tag{6.5}$$

It depends on the main and the angular momentum quantum numbers n and l. Here A denotes the normalization constant and Z is the nuclear charge. The formula includes two empirical parameters, the effective main quantum number n^* and the screening constant s.   **screening constant**
The angular dependence of the effective one-particle wavefunctions are identical to those of the hydrogen atom, the spherical harmonics Y_{lm}. The approximate one-electron wavefunctions of an atom or ion are then given by

$$\psi_{nlm}(\rho, \vartheta, \varphi) = R_{nl}(\rho)\,Y_{lm}(\vartheta, \varphi) \tag{6.6}$$

and are called Slater orbitals.   **Slater orbitals**

The total energy E_{tot} of an atom or ion – the negative of the energy required to remove all the electrons from the nucleus to an infinite distance – is given by the sum of the orbital energies ε_{nl} of all N electrons:

$$E_{tot} = \sum_{n,l}^{N} N_{nl}\varepsilon_{nl} \quad \text{with} \quad \varepsilon_{nl} = -\frac{1}{2}\left(\frac{Z-s}{n^*}\right)^2 \tag{6.7}$$

where N_{nl} denotes the number of electrons in a given state with main angular momentum quantum numbers n and l, and energies are given in units of Hartree. Comparing this result with the energy levels of the one-electron problem in a central field (i.e., the hydrogen atom), we realize that the quantity $Z_{eff} = Z - s$ is the mean effective charge. The Slater rules are a recipe to determine the parameters s and n^*, which will immediately allow evaluation of the one-electron energies. These can then be used to calculate total energies and related quantities, for example, ionization potentials. For the calculation of atomic or ionic radii, the maximum of the radial probability density $4\pi\rho^2(R_{nl}(\rho))^2$ has to be calculated, which is found at

$$\rho_{max} = \frac{(n^*)^2}{Z - s} \tag{6.8}$$

In his seminal paper, Slater used his rules to calculate the energy levels and total energies of ions and atoms, ionization potentials, and extensions of atoms and ions. The following passage is taken from Slater's 1930 paper [2]:

The values for n^* and $Z - s$ are given by the following rules:

1. n^* is assigned by the following table, in terms of the real principal quantum number n:

n	1	2	3	4	5	6
n^*	1	2	3	3.7	4.0	4.2

2. For determining $Z - s$, the electrons are divided into the following groups, each having a different constant: *(1s); (2s,2p); (3s,3p); (3d); (4s,4p); (4d); (4f); (5s,5p); (5d)*; etc. That is, the s and p electrons of a given n are grouped together, but the d and f are kept separated. The shells are considered to be arranged from inside out in the order named.

3. The shielding constant s is formed, for any group of electrons, from the following contributions:
   a) Nothing from any shell outside the one considered.
   b) An amount 0.35 from each other electron in the group considered (except in the 1s group, where 0.30 is used instead).
   c) If the shell considered is an (s,p) shell, an amount of 0.85 from each electron with total quantum number less by one, and an amount of 1.00 from each electron still further in; but if the shell is a d or f, an amount of 1.00 from every electron inside it.

Table 6.1 Screening constants s according to the Slater rules.

Considered electron	Screening electrons							
	1s	2s	2p	3s	3p	3d	4s	4p
1s	0.3	0	0	0	0	0	0	0
2s, 2p	0.85	0.35	0.35	0	0	0	0	0
3s, 3p	1.0	0.85	0.85	0.35	0.35	0	0	0
3d	1.0	1.0	1.0	1.0	1.0	0.35	0	0
4s, 4p	1.0	1.0	1.0	0.85	0.85	0.85	0.35	0.35

The screening constants according to these rules are summarized in Table 6.1.

6.2.3
Calculation of Ionization Potentials and Electron Affinities

Independent of the employed method, we can always calculate the IP and the EA using Equations 6.1 and 6.4, respectively. For the ionization potential it is also possible to use Koopmans' theorem (Equation 6.2). Koopmans' theorem looks more appealing than Equation 6.1, as it involves only a single calculation. Koopmans showed that, within Hartree–Fock theory and for closed-shell systems, Equation 6.2 is a good approximation for "electron detachment energies," which are equivalent to IPs.

There are, however, various reasons to suggest that one should use Equation 6.1 instead. First, there is the physical picture suggesting that, after release of an electron, the electronic structure of the remaining ion will relax, and hence the system will change its energy. Therefore IPs will be overestimated using Koopmans' theorem. Second, the orbital energies are not necessarily well defined in the quantum method that we employ for the approximate solution of the Schrödinger equation. This concern affects, in particular, density-functional theory. Third, there is a technical issue: various numerical details such as basis-set size and so on will affect the energy value of the HOMO. By calculating the difference between the total energies of the molecule and the ion, many of those systematic errors will cancel, and the numerical values will be more accurate. These three reasons indicate that it is usually not a good idea to employ the analog of Koopmans' theorem to calculate EAs, which would be the negative of the energy of the lowest unoccupied molecular orbital (LUMO).

Technically, it is important to note that standard methods of quantum chemistry are designed to describe neutral atoms. This means that basis sets cover comfortably the electrons of atoms, and those of cations.

LUMO

For negatively charged species, the situation is different, as the additional negative charge results in spatial extension of the anions, which might not be covered by standard basis sets.

Consistently with experiments, IPs and EAs should be given in units of electronvolts (eV), the conversions being 1 Hartree = 27.211 eV, and 1 eV = 96.485 kJ mol^{-1}.

6.3
Demonstration

We will calculate the IPs and EAs of the potassium and argon atoms using the methods described above and compare the results with experiment. IPs and EAs are available in the literature. Beside the standards, the web page http://www.webelements.com is an excellent source of information. There we find the numerical values listed in Table 6.2 (due to its unstable anion, the EA of Ar is not defined).

Table 6.2 Experimental IP and EA values for potassium and argon.

Element	Electron configuration	IP (eV)	EA (eV)
K	$1s^2$, $2s^2$, $2p^6$, $3s^2$, $3p^6$, $4s^1$	4.34	0.51
Ar	$1s^2$, $2s^2$, $2p^6$, $3s^2$, $3p^6$	15.76	—

Following the Slater rules we first calculate the shieldings and effective nuclear charges and energies of each of the various groups of electrons. These are given in Table 6.3.

Following Koopmans' theorem, we obtain IPs of 0.177 Hartree = 4.81 eV for K, and 2.53 Hartree = 68.84 eV for Ar. If we compare these results with experiment (experimental values in Table 6.2), we can see that the IP for K has acceptable accuracy, taking into account the crude level of computation, but the result for Ar is off by a factor of 4! What is the reason for this error?

Let us calculate the IP using Equation 6.1, by the difference between the total energies of the atom and ion. Within Slater's concept, the total energy is the sum of electron energies:

$$E_{tot} = \sum_{nl} n_{nl}\varepsilon_{nl} \qquad (6.9)$$

The sum runs over all occupied shells, and the occupation number N_{nl} gives the number of electrons in each shell. Application of Equation 6.9 leads to

$$E_{tot}(K) = 2\varepsilon_{1s} + 8\varepsilon_{2s,2p} + 8\varepsilon_{3s,3p} + 1\varepsilon_{4s} = -597.083 \text{ Hartree}$$

Table 6.3 Calculation of the shieldings of potassium and argon following Slater's rules.

Electrons		K	
	s	Z_{eff}	ε (Hartree)
1s	$1 \times 0.30 = 0.30$	$19 - 0.30 = 18.70$	$-0.5 \times (18.7/1)^2 = -174.845$
2s,2p	$2 \times 0.85 + 7 \times 0.35 = 4.15$	$19 - 4.15 = 14.85$	$-0.5 \times (14.85/2)^2 = -27.565$
3s,3p	$2 \times 1.00 + 8 \times 0.85 + 7 \times 0.35 = 11.25$	$19 - 11.25 = 7.75$	$-0.5 \times (7.75/3)^2 = -3.337$
4s	$2 \times 1.00 + 8 \times 1.00 + 8 \times 0.85 = 16.80$	$19 - 16.80 = 2.20$	$-0.5 \times (2.2/3.7)^2 = -0.177$

Electrons		Ar	
	s	Z_{eff}	ε (Hartree)
1s	$1 \times 0.30 = 0.30$	$18 - 0.30 = 17.70$	$-0.5 \times (17.7/1)^2 = -156.645$
2s,2p	$2 \times 0.85 + 7 \times 0.35 = 4.15$	$18 - 4.15 = 13.85$	$-0.5 \times (13.85/2)^2 = -23.978$
3s,3p	$2 \times 1.00 + 8 \times 0.85 + 7 \times 0.35 = 11.25$	$18 - 11.25 = 6.75$	$-0.5 \times (6.75/3)^2 = -2.531$
4s	—	—	—

$$E_{tot}(K^+) = 2\varepsilon_{1s} + 8\varepsilon_{2s,2p} + 8\varepsilon_{3s,3p} = -596.906 \text{ Hartree}$$

$$E_{tot}(Ar) = 2\varepsilon_{1s} + 8\varepsilon_{2s,2p} + 8\varepsilon_{3s,3p} = -525.434 \text{ Hartree}$$

For the calculation of the total energy of Ar^+ we need to take into account that an electron has to be removed from the 3s,3p shell. According to the Slater rules, the remaining seven electrons are shielded less compared to those in the Ar atom. We obtain

$$s_{3s,3p} = 2 \times 1.00 + 8 \times 0.85 + 6 \times 0.35 = 10.9$$

$$\varepsilon_{3s,3p} = -0.5 \times (7.1/3)^2 = -2.801 \text{ Hartree}$$

$$E_{tot}(Ar^+) = 2\varepsilon_{1s} + 8\varepsilon_{2s,2p} + 7\varepsilon_{3s,3p} = -524.793 \text{ Hartree}$$

The result does not change for the IP of K. Removing the only electron of the 4s shell does not affect the lower shells, and Koopmans' theorem gives the identical result. Note that one should expect the very same behavior for any method that treats only valence electrons, for example, most semi-empirical methods or *ab initio* methods using pseudo-potentials to include the core electrons. The situation is different for Ar. As one electron is removed from the 3s,3p shell, the remaining electrons in this shell are shielded less and hence are subject to a higher effective charge Z_{eff}, and the IP is strongly reduced compared to the prediction by Koopmans' theorem, with a

Table 6.4 Results of density-functional calculations with deMon (PBE/A2*/DZVP-GGA). All values are given in Hartree, if not stated otherwise.

	K	Ar
ε_{HOMO}	−0.09185	−0.37723
$E_{tot}(X)$	−599.472997	−527.142143
$E_{tot}(X^+)$	−599.309887	−526.561964
$E_{tot}(X^-)$	−599.486948	−526.621451
IP (eV), Equation 6.2	2.50	10.26
IP (eV), Equation 6.1	4.44	15.78
IP (eV), experiment	4.34	15.76
EA (eV), Equation 6.4	0.37	−14.17
EA (eV), experiment	0.50	—

numerical value of 17.44 eV, again in acceptable agreement with the experimental value of 15.76 eV.

Let us perform a more sophisticated calculation of the IPs and EAs by solving numerically the approximate Schrödinger equation. To be on the safe side, we should use numerical details such as a comfortable basis set for these calculations, in particular to compute EAs. The calculation yields the results presented in Table 6.4.

Obviously, Koopmans' theorem is not a good approximation for our way to solve the approximate Schrödinger equation (which is density-functional theory). However, calculating ionization potentials using Equation 6.1 works very well, the difference from experiment being only ∼0.1 eV. Electron affinities are generally harder to calculate, and here we can – at best – hope to get acceptable trends without further tweaking our numerical method. Our calculation correctly predicts that Ar⁻ is unstable, and a small value for the electron affinity of K.

6.4
Problems

1. The ionization potentials of the second period

Determine the first IPs of the elements of the second period:
a) using the Slater rules and Equation 6.1;
b) using the Slater rules and Equation 6.2; and
c) by numerical solution of the approximate Schrödinger equation and Equation 6.1.

Compare your results with data taken from the literature, for example, from the WebElements homepage (http://www.webelements.

com), by plotting them in a diagram (IP versus atomic number, a common diagram found in textbooks).

2. The electron affinities of the second period

Determine the EAs of the elements of the second period using Equation 6.4, taking total energies from numerical solutions of the Schrödinger equation. Compare the results with experiment (see problem 1) and plot them in a diagram (EA versus atomic number).

3. Quality of theoretical predictions – comparison to experiment

Assuming that the experimental values are the reference, compare the quality of your theoretical predictions. Make a scatterplot of calculated versus experimental data and calculate the best-fit straight line. Calculate the root-mean-square deviation from experiment for each method. Which conclusions can be drawn concerning the quality of your calculated results?

4. IPs and EAs for the first, fourth, and seventh main groups

Calculate the IPs and EAs for the first four shells of the main group elements of the first main group, fourth main group, and seventh main groups using the same approach as in problem 1c.

5. Interpret your results

a) What is the general trend of the IPs and EAs going to higher atomic numbers in one main group?
b) What is the general trend for IPs going to higher atomic numbers in one period?
c) The first IP of B is lower than that of Be. Explain this exception from the general trend.

6.5
Technical Details

For all electron structure calculations, you can use deMon as provided on the CD delivered with this book. The input file for the He atom is given below:

```
TITLE HE ATOM
VXCTYPE AUXIS PBE
BASIS (DZVP-GGA)
AUXIS (GEN-A2*)
GEOMETRY CARTESIAN
HE  0.0  0.0  0.0
END
```

Additional keywords must be provided:

```
CHARGE 1
```

for the cation, and

```
CHARGE -1
```

for the anion.

In case the calculation does not converge use additionally the keyword `SHIFT 0.2`

6.6
Review and Summary

The physical grounds of the definition of the ionization potential and the electron affinity have been introduced. First, the approach to compute ionization potentials through the Slater rules has been discussed. Though empirical, it is a very useful as very illustrative approach. Then, IPs and EAs have been calculated using modern computational chemistry. While the first approach shows large error bars, the latter approach gives reasonable values for IPs and also for most EAs.

Furthermore, we have learned that EAs might not be well defined, as the anion might not always be a stable system. Finally, you have learned to compare the energies of two systems. The main contribution of the total energy, in particular for heavier atoms, is caused by the core electrons, and total energy differences of two systems are small compared to the values of the total energies themselves.

References

1 Koopmans, T.A. (1934) Über die Zuordnung von Wellenfunktionen und Eigenwerten zu den einzelnen Elektronen eines Atoms. *Physica*, **1**, 104–113.

2 Slater, J.C. (1930) Atomic shielding constants. *Physical Review*, **36**, 57–64.

7
Hückel Molecular Orbital Theory:
Stability of Conjugated Carbon π Systems

7.1
Aim

Hückel theory is a powerful tool to study conjugated carbon π systems. **conjugated carbon π system**
It is also an excellent way to introduce the linear combination of atomic
orbitals (LCAO) concept, which is the backbone of most quantum
chemical computer codes. We will study carbon π systems using Hückel
molecular orbital (HMO) theory. We will introduce the LCAO concept
with acceptable mathematical effort, and show how an approximate
solution of the Schrödinger equation of a model system can be obtained.
The resulting molecular orbital energy levels and binding energies
are used to discuss the stability of annulenes and fullerenes. The latter
example shows that even nowadays simple concepts can be used to
predict structures and to understand experimental results.

7.2
Theoretical Background

Hückel molecular orbital theory is a simplified version of molecular **Hückel molecular orbital**
orbital (MO) theory and applicable for planar (or, with some approxi- **theory**
mation, nearly planar) molecules, which consist of sp^2 hybridized
carbon atoms and, therefore, have an extended π electron system. It
is a very simple, but effective, semi-empirical method that was developed in 1931 by the German physicist–physicochemist Erich Hückel
(1896–1980) to perform calculations on conjugated hydrocarbons. It
has been used, for example, to establish the Hückel rules for
annulenes, to understand the color of dyes, and to predict the stability
and electronic properties of nanostructures, for example, fullerenes **fullerene**
and nanotubes. There are extensions to Hückel theory not covered in
this workbook, such as for example extended Hückel theory (EHT),
which is applicable to any molecule or material and which is known
under the term tight-binding method in physics and materials science.

Computational Chemistry Workbook: Learning Through Examples
Thomas Heine, Jan-Ole Joswig, and Achim Gelessus
Copyright © 2009 WILEY-VCH Verlag GmbH & Co. KGaA, Weinheim
ISBN: 978-3-527-32442-2

7.2.1
Molecular Orbital Theory

The total energy of a molecule is the sum of the electronic energy and nucleus–nucleus repulsion:

$$E_{tot} = E_{el} + E_{nuc\text{-}nuc}$$

In order to calculate the electronic energy E_{el}, the wavefunction Ψ or various other properties of an atom or molecule quantum mechanically, the electronic stationary Schrödinger equation

$$\hat{H}\Psi = E_{el}\Psi \tag{7.1}$$

needs to be solved. This is a second-order partial differential equation containing $3N$ variables (N being the number of electrons). The most common way to solve such a highly dimensional partial differential **product ansatz** equation is the product ansatz, that is, the solution is a product of functions, each depending on fewer independent variables. Therefore, in molecular orbital theory we write the wavefunction Ψ as the product of functions (or orbitals) ψ_i, which depend on the three spatial positions of the electron in the particular orbital:

$$\Psi(x_1, y_1, z_1, x_2, y_2, z_2, \ldots, x_N, y_N, z_N)$$
$$= \psi_1(x_1, y_1, z_1) \times \psi_2(x_2, y_2, z_2) \times \cdots \times \psi_N(x_N, y_N, z_N) \tag{7.2}$$

separation of variables   Through this separation of variables, the single Schrödinger equation, Equation 7.1, depending on $3N$ variables, is split into N single-particle equations,

$$\hat{h}\psi_i = \varepsilon_i\psi_i \tag{7.3}$$

with $i = 1, 2, \ldots, N$, each depending on only three variables. The coordinate index can be omitted, and each orbital ψ_i is a function of the spatial coordinates (x, y, z). The Hamiltonian $\hat{h}$ is an operator that has the same mathematical form for all orbitals, and $\hat{h}$ acts on each orbital, which can host a single electron, individually. All other electrons are present in $\hat{h}$ through a parameterized effective mean potential. As the electron–electron interactions, which are implicitly present in the Hamiltonian, are neglected, the electronic energy results in

$$E_{el} = \sum_{i=1}^{N} \varepsilon_i \tag{7.4}$$

The variables ε_i and ψ_i are now the energies and the wavefunctions of the resulting N molecular orbitals. In the single-particle equations (Equations 7.3), the functions ψ_i are expanded in a linear combination of n atomic orbitals (AOs) φ_j. These are atomic functions that are known, for example, from solving the Schrödinger equation for the free atom:

$$\psi_i = \sum_{j=1}^{n} c_{ij}\varphi_j \tag{7.5}$$

This ansatz is called linear combination of atomic orbitals (LCAO). As a result, the unknown parameters are no longer the functions ψ_i, but the coefficients c_{ij}, the so-called MO coefficients. In order to optimize the set of MO coefficients with respect to the electronic energy, we have to make use of the variational principle. This leads to the following set of linear equations: **linear combination of atomic orbitals (LCAO)**

molecular orbital variational principle

$$\sum_{j=1}^{N} (h_{ij} - \varepsilon_i s_{ij}) c_{ij} = 0 \tag{7.6}$$

Here, h_{ij} and s_{ij} are the Hamilton integral and overlap integral, respectively, containing the atomic orbitals: **Hamilton integral** **overlap integral**

$$h_{ij} = \langle \varphi_i | \hat{h} | \varphi_j \rangle \quad \text{and} \quad s_{ij} = \langle \varphi_i | \varphi_j \rangle$$

The set of linear equations has solutions if the determinant is zero, that is,

$$\det \begin{pmatrix} h_{11} - \varepsilon_1 s_{11} & h_{12} - \varepsilon_1 s_{12} & \cdots & h_{1n} - \varepsilon_1 s_{1n} \\ h_{21} - \varepsilon_2 s_{21} & h_{22} - \varepsilon_2 s_{22} & \cdots & h_{2n} - \varepsilon_2 s_{2n} \\ \vdots & \vdots & \ddots & \vdots \\ h_{n1} - \varepsilon_n s_{n1} & h_{n2} - \varepsilon_n s_{n2} & \cdots & h_{nn} - \varepsilon_n s_{nn} \end{pmatrix} = 0 \tag{7.7}$$

This leads to a polynomial of order n, which results in n orbital energies ε_i. These are the eigenvalues of the eigenvalue problem given above (Equation 7.6). With Equation 7.6, these n eigenvalues lead to a set of n^2 MO coefficients c_{ij} of the eigenvectors c_i. The MO coefficients give the magnitude of each AO contribution to the molecular orbital. In a closed-shell system, the energetically lowest $N/2$ MOs are doubly occupied and called occupied orbitals. The unoccupied orbitals are also called virtual orbitals. **MO coefficients** **MO eigenvectors**

virtual orbitals

So far, we have implicitly used the Born–Oppenheimer approximation to obtain the stationary electronic Schrödinger equation, the separation of variables (product ansatz) to obtain one-particle equations, and the LCAO ansatz to solve the set of one-particle equations. Moreover, we have approximated the electron–electron interactions to be described by an average potential. The problem is now strongly simplified, but still requires the calculation of numerous overlap matrix elements and Hamilton matrix elements.

Simply speaking, the Born-Oppenheimer approximation assumes that the electronic and nuclear motion are independent due to the large difference in mass between the electron and the nucleus. The Born-Oppenheimer approximation will be discussed more detailed in chapters 9 and 15.

Further approximations refer to the still three-dimensional Hamilton and overlap integrals. The Hamiltonian matrix elements are now approximated as empirical parameters that contain, for example, electron–electron and nucleus–nucleus interactions. All simplifications and approximations within Hückel theory will be summarized in the following.

7.2.2
The Hückel Postulates

Hückel postulates

Although HMO theory [1] uses many approximations and simplifications, the resulting description of the π orbitals is qualitatively, and sometimes even quantitatively, quite accurate. The basis for HMO theory is formed by a number of Hückel postulates, which will be listed here first and then exemplified below: (1) σ/π separation, (2) LCAO approximation, (3) neglect of orbital overlap resulting in orthonormal basis functions, and (4) approximation of the interaction potential present in the Hamilton integrals. The details of these approximations are as follows.

σ/π separation

1. **σ/π separation:** The σ and π electrons of the total system will be treated separately and independently. This is an important approximation, because the statements made through Hückel theory refer to the π system of the molecule only. In practice, the σ electrons are neglected. This approximation assumes the presence of π electrons, so it can only be applied to systems where σ and π electrons can be clearly separated. Besides planar molecules, these systems may include, for example, fullerenes, carbon nanotubes and other carbon nanostructures, for which HMO theory is still popular.

2. **LCAO approximation:** To describe the molecular orbitals, the LCAO ansatz is used, but only one orbital per atom is considered. The MO wavefunction ψ_i of the ith orbital of a molecule containing N sp^2 hybridized carbon atoms will, therefore, be written as

$$\psi_i = \sum_{j=1}^{N} c_{ij}\varphi_j$$

Each molecular orbital thus consists of contributions of N atomic orbitals. Because of the σ/π separation, only one atomic p orbital per carbon atom (perpendicular to the molecular plane) is considered. The coefficients c_{ij} determine the contribution of the jth atomic orbital to the ith molecular orbital. The coefficients are real numbers and may be positive or negative. If $c_{ij} = 0$, the wavefunction has a node at the corresponding carbon atom.

3. **Neglect of orbital overlap – orthonormal basis functions**: All overlap interactions of non-neighboring molecules are neglected. Moreover, the overlap matrix elements describing next-neighbor interactions are set to zero and considered in the parameterization of the Hamilton matrix elements. The only non-zero overlap matrix elements are those on the diagonal, which are normalized by definition:

$$S_{kl} = \int_\tau \varphi_k^* \varphi_l \, d\tau = \langle k|l \rangle = \delta_{kl}$$

Therefore, the overlap matrix is the identity matrix, $S = 1 = (\delta_{kl})$, that is, $S_{kk} = 1$ and $S_{kl} = 0$ (for $k \neq l$).

4. **Approximation of the interaction integrals**: Instead of calculating the Hamiltonian matrix elements $\langle \varphi_k | \hat{H} | \varphi_l \rangle$ exactly, they will be approximated as follows.

   a) **On-site Hamiltonian matrix element**: The interaction integral of a π electron in the atomic orbital φ_k at atom k is defined as

   $$\langle \varphi_k | \hat{H} | \varphi_k \rangle = H_{kk} = \alpha$$

   A reasonable value for this integral is the energy of a π electron in the p orbital of the free atom. These integrals are sometimes called (ambiguously) Coulomb integrals.   **Coulomb integral**

   b) **Off-site Hamiltonian matrix elements**: The interaction between two π electrons in the atomic orbitals φ_k and φ_l at different atoms k and l is defined as

   $$\langle \varphi_k | \hat{H} | \varphi_l \rangle = H_{kl} = \begin{cases} \beta & (\text{if } k \text{ and } l \text{ are neighbours}) \\ 0 & (\text{if } k \text{ and } l \text{ are not neighbours}) \end{cases}$$

   The value for β corresponds to the energy of a π electron in the field of two nuclei. These integrals are sometimes called resonance integrals.   **resonance integral**

7.2.3
Topology Matrices

Besides these four approximations, the problem can be simplified through the assumption that all sp^2 carbon atoms are equivalent. We have already made use of this by setting all integrals to either α or β and neglecting the atomic indices. The resulting secular determinant (Equation 7.7) is then

$$\begin{vmatrix} \alpha - \varepsilon & \beta & \cdots & \beta \\ \beta & \alpha - \varepsilon & \cdots & \beta \\ \vdots & \vdots & \ddots & \vdots \\ \beta & \beta & \cdots & \alpha - \varepsilon \end{vmatrix} = 0 \qquad (7.8)$$

Note that this is only an example, since the resonance integrals, which are the off-diagonal elements in the matrix above, can adopt the values β or 0 depending on the connectivity of the participating atoms. Dividing by β and with $x = (\alpha - \varepsilon)/\beta$, we obtain

$$
\begin{vmatrix}
(\alpha - \varepsilon)/\beta & 1 & \cdots & 1 \\
1 & (\alpha - \varepsilon)/\beta & \cdots & 1 \\
\vdots & \vdots & \ddots & \vdots \\
1 & 1 & \cdots & (\alpha - \varepsilon)/\beta
\end{vmatrix}
=
\begin{vmatrix}
x & 1 & \cdots & 1 \\
1 & x & \cdots & 1 \\
\vdots & \vdots & \ddots & \vdots \\
1 & 1 & \cdots & x
\end{vmatrix}
= 0
$$

(7.9)

Hückel topology matrix   If we subtract the diagonal matrix $x\delta_{kl}$, we obtain the Hückel topology matrix $\mathbb{A}$ (sometimes also called adjacency matrix)

$$
\begin{pmatrix}
0 & 1 & \cdots & 1 \\
1 & 0 & \cdots & 1 \\
\vdots & \vdots & \ddots & \vdots \\
1 & 1 & \cdots & 0
\end{pmatrix}
$$

of the sp^2 carbon atoms of the molecule: the matrix element $\mathbb{A}_{kl}$ is 1 if k and l are next neighbors and 0 otherwise, which means that it gives the connectivity pattern of the conjugated molecule. Now we can see that only the adjacency pattern of a molecule enters the Hamiltonian, and hence the result, in HMO theory. It is only necessary to know the connectivity of a molecule – detailed knowledge of bond lengths and so on is not necessary.

In consequence, the topology matrix is most conveniently used to start a Hückel calculation, as it only differs from the Hückel matrix by the known diagonal $x\delta_{kl}$. As an example, the Hückel matrix for 1-methylene-cyclopropene is

$$
\det \boldsymbol{H} = \det
\begin{pmatrix}
x & 1 & 1 & 0 \\
1 & x & 1 & 0 \\
1 & 1 & x & 1 \\
0 & 0 & 1 & x
\end{pmatrix}
= 0
$$

(7.10)

All diagonal elements are set to x, whereas all off-diagonal elements are set to either 0 or 1 (corresponding to either 0 or β). That is, if atoms j and k are connected, the Hückel matrix element H_{jk} is set to 1, otherwise it is 0. Setting the determinant to zero and solving this set of linear equations results in eigenvalues x_i and eigenvector components c_{ij} of the eigenvectors c_i. To obtain the energies, a re-substitution according to

$$
\varepsilon_i = \alpha - x_i\beta
$$

(7.11)

has to be made.

7.2.4
Values for α and β

At first glance, using only the constants α and β might be confusing, but hard numbers are actually not relevant in Hückel theory for systems that contain only carbon (and Hückel-irrelevant hydrogen) atoms as far as binding energies are concerned. The resulting energies are given in the form of Equation 7.11 with the only varying parameter being the x_i. This means that the energies are given with respect to the orbital energy of the free carbons, α, and should be understood as orbital stabilization or destabilization (if ε_i is smaller or larger than α, respectively). For this reason, it is not necessary to know a numerical value for α. The constant β is nothing other than the unit of our energy scale, which does not need to be a real number for qualitative considerations.

Nevertheless, for the calculation of orbital differences, for example, to estimate the HOMO–LUMO gap, a numerical value is needed. The choice $\beta = -0.1$ Hartree is a good approximation for this purpose. Note that the strongly approximate character of Hückel theory implies that quantitative results should be interpreted very cautiously.

7.2.5
The Trap of Defining x

There are two ways of defining x, which unfortunately are both used throughout the literature and lecture notes. However, since it is a substitution, one only has to be aware of the problem and re-substitute in the right way, as is shown in Table 7.1 for the cyclopropenyl system.

Table 7.1 Two ways of substituting x for the cyclopropenyl system. Both result in the same eigenvalues.

	$+x$	$-x$
Substitution	$x = \dfrac{\alpha - \varepsilon}{\beta}$	$-x = \dfrac{\alpha - \varepsilon}{\beta}$
Re-substitution	$\varepsilon = \alpha - x\beta$	$\varepsilon = \alpha + x\beta$
Determinant	$\begin{vmatrix} x & 1 & 1 \\ 1 & x & 1 \\ 1 & 1 & x \end{vmatrix} = 0$ $\Rightarrow x^3 - 3x + 2 = 0$ $\Leftrightarrow (x-1)(x-1)(x+2) = 0$	$\begin{vmatrix} -x & 1 & 1 \\ 1 & -x & 1 \\ 1 & 1 & -x \end{vmatrix} = 0$ $\Rightarrow -x^3 + 3x + 2 = 0$ $\Leftrightarrow (x+1)(x+1)(x-2) = 0$
Results and re-substitution	$x_1 = +1$ $\quad \varepsilon_1 = \alpha - \beta$ $x_2 = +1 \Rightarrow \varepsilon_2 = \alpha - \beta$ $x_3 = -2 \quad \varepsilon_3 = \alpha + 2\beta$	$x_1 = -1$ $\quad \varepsilon_1 = \alpha - \beta$ $x_2 = -1 \Rightarrow \varepsilon_2 = \alpha - \beta$ $x_3 = +2 \quad \varepsilon_3 = \alpha + 2\beta$

Both substitutions lead to the same energy eigenvalues. If you are not sure which substitution pattern is used in a computer program, look at the MO coefficients of the orbitals. The orbital with no nodes, that is, where all the MO coefficients have the same sign, always corresponds to the lowest energy orbital.

7.2.6
π Electron Binding Energy

π electron energy The π electron energy is defined as

$$E_\pi = \sum_{i=1}^{N} n_i \varepsilon_i$$

Here, n_i is the occupation number of orbital i, ε_i its orbital energy, and N the number of carbon atoms. On the Hückel level, the binding energy E_B is obtained by taking the difference between the π electron energy, E_π, and the energy of the free carbon atom, $E_C = \alpha$:

$$E_B = N E_C - E_\pi = N\alpha - \sum_{i=1}^{N} n_i \varepsilon_i = \beta \sum_{i=1}^{N} n_i x_i \qquad (7.12)$$

In this way the parameter α is eliminated, and the binding energy gives the energy relative to the energy of the free carbon atoms.

7.2.7
π Electron Molecular Orbitals and Probability Density

probability density The probability density of a p electron in the ith molecular orbital is given by the standard representation: $\rho_i = |\psi_i|^2 = \psi_i \psi_i^*$. Within the LCAO scheme and because of the fact that the MO coefficients and atomic orbitals are real numbers, we obtain

$$\rho_i = \psi_i \psi_i^* = \sum_{kl} c_{ik} \varphi_k c_{il}^* \varphi_l^* = \sum_{kl} c_{ik} c_{il} \varphi_k \varphi_l$$

If we also use the Hückel approximation that the atomic orbitals do not overlap, we can for a qualitative approximation assume that $\varphi_k \varphi_l \approx 0$ for $k \neq l$ and obtain

$$\rho_i \approx \sum_{k} c_{ik}^2 \varphi_k^2$$

molecular orbital The form of a molecular orbital is determined by its nodal structure. As atomic orbitals are normalized, it is a good approximation to sketch a p-based π MO in the following way:

1. Plot the framework of the molecule (without hydrogen and double bonds).

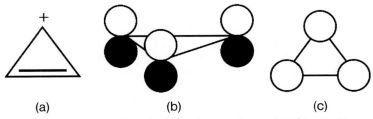

Figure 7.1 (a) The chemical formula of the cyclopropenyl cation $C_3H_3^+$ in one of the possible mesomeric structures (hydrogen atoms are not shown). (b) The three p orbitals of the cyclopropenyl cation that are relevant for the Hückel calculation (side view) and (c) the same three p orbitals (top view).

2. Calculate the square of the MO coefficients of the orbital.
3. Draw circles with radii of the square of the corresponding MO coefficient around each carbon.
4. Indicate the nodes – fill all circles that correspond to negative MO coefficients in black.

This procedure will be followed in the next section, and an example is given in Figure 7.3. Remember that the plotted AO contributions are p orbital contributions. Since we most often look at planar carbon systems from a top view, we only see one (spherical) part of the dumbbell-shaped p orbitals (see Figure 7.1).

7.3
Demonstration

7.3.1
Hückel Calculation of the Cyclopropenyl Cation

As an example, we will examine the cyclopropenyl cation. Figure 7.1a shows the chemical formula that is usually used in organic chemistry. Moreover, a side view (Figure 7.1b) and a top view (Figure 7.1c) of the Hückel-relevant carbon backbone (without hydrogen atoms) is shown, including the three p orbitals. It is most important to bear in mind that all the orbitals discussed in HMO theory originate from p orbitals and have contributions below and above the molecular plane.

To set up the topology (or adjacency) matrix it is only necessary to know which carbon atoms are bonded to each other. It is not necessary to know, or to guess, the location of double bonds. Hydrogen atoms are ignored. It is, however, necessary to check that all the carbon atoms entering the Hückel calculations are trivalent (sp^2 carbons). So in our example each carbon atom has a hydrogen atom attached. Also, the location of the charge is irrelevant at this point, except for the fact that it

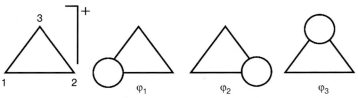

Figure 7.2 Structure of cyclopropenyl cation (left) and its three atomic orbitals φ_k that enter the LCAO calculation (right). The atom numbering is also shown.

changes the number of π electrons. The input structure is as shown in Figure 7.2.

The right-hand side of Figure 7.2 shows schematically the three atomic p orbitals that form the basis of the calculation. The topology matrix is very simple, because each atom is connected to all other atoms, and the Hückel matrix is given by

$$H = \begin{pmatrix} x & 1 & 1 \\ 1 & x & 1 \\ 1 & 1 & x \end{pmatrix} \quad \text{with} \quad x = \frac{\alpha - \varepsilon}{\beta} \qquad (7.13)$$

The resulting eigenvalue problem is particularly simple and can be solved analytically. To obtain the eigenvalues, a polynomial of third order has to be solved:

$$x^3 - 3x + 2 = 0 \quad \Leftrightarrow \quad (x-1)(x-1)(x+2) = 0$$

giving

$$\begin{aligned} x_1 &= +1 & \varepsilon_1 &= \alpha - \beta \\ x_2 &= +1 & \Rightarrow \quad \varepsilon_2 &= \alpha - \beta \\ x_3 &= -2 & \varepsilon_3 &= \alpha + 2\beta \end{aligned}$$

The eigenvalues can be visualized schematically in an occupation scheme, as shown in Figure 7.3. After obtaining the three eigenvalues, we now have to calculate the three eigenvectors. For small systems, this can be done by hand; for larger systems, this can always be done with computer programs. The eigenvectors need to be normalized to fulfill the basic requirements of quantum mechanics. The resulting eigenvectors could be

$$c_1 = \frac{1}{\sqrt{2}} \begin{pmatrix} +1 \\ -1 \\ 0 \end{pmatrix}, \quad c_2 = \frac{1}{\sqrt{6}} \begin{pmatrix} +1 \\ +1 \\ -2 \end{pmatrix}, \quad c_3 = \frac{1}{\sqrt{3}} \begin{pmatrix} +1 \\ +1 \\ +1 \end{pmatrix}$$

As there are only two π electrons in the cyclopropenyl cation, both go into the lowest energy orbital ψ_3 with energy ε_3 (note that $\beta < 0$), which is

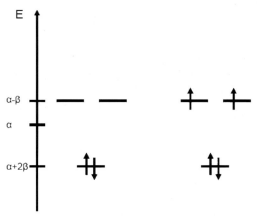

Figure 7.3 Eigenvalues and occupation scheme of the cyclopropenyl cation (left) and the cyclopropenyl anion (right).

characterized by the MO coefficients of c_3. The MO coefficients indicate a symmetric shape of the ψ_3 MO. It is sketched in Figure 7.4a.

Note that the two high-energy orbitals ε_1 and ε_2 have the same energy – we speak of degenerate orbitals. This means that there is no preference in their occupation. If we had performed calculations on the cyclopropenyl anion, which has four π electrons, following Hund's rule both orbitals would be occupied by a single electron each. This would lead to a radical, which actually indicates that this anion is unstable (this is confirmed by the fact that those orbitals are antibonding, that is, have

degenerate orbital

Hund's rule

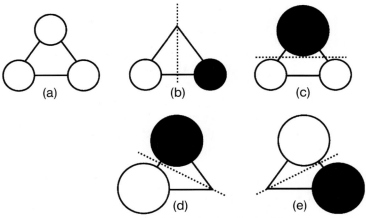

Figure 7.4 Schematic visualization of the molecular orbitals of the cyclopropenyl cation. The lowest energy orbital (a) and the two high-energy orbitals (b and c) are indicated. An alternative representation of the degenerate high-energy orbitals is sketched (d and e). Nodes are indicated as dotted lines.

a negative binding energy). The shape of the two orbitals is given in Figure 7.4b and c. Note the node in each orbital, which goes either through an atom or through the bonds. Remember also that these are π orbitals with a nodal plane in the paper.

A second note is necessary to deal with degenerate orbitals. Degenerate orbitals have to be treated on an equal footing. This means that you cannot tell if c_1 or c_2 belongs to either ε_1 or ε_2. In fact, any orthonormalized linear combination of c_1 and c_2 satisfies the linear system of equations given by Equations 7.13. You can try to build a different set of solutions, for example, by forming two new orbitals:

$$c'_1 = \frac{1}{\sqrt{2}}(c_1 + c_2), \quad c'_2 = \frac{1}{\sqrt{2}}(c_1 - c_2), \quad c'_3 = c_3$$

Note that the choice of the first orbital determines the choice of the second, because orthonormalization is required. The meaning of this transformation may be understood in the easiest way if you compare the shapes of the new MOs, which are given in Figure 7.4d and e.

7.4
Problems

1. Hückel's rule for aromaticity

a) Calculate the Hückel energy eigenvalues of the [n]-annulenes C_nH_n ($n = 4, 6, 8, 10$) in D_{nh} symmetry.

b) Draw the eigenvalue spectra of each molecule and fill in the appropriate number of π electrons. Deduce Hückel's rule for aromaticity from your drawings.

2. C_{76} fullerenes

There are two isomers for reasonably structured C_{76} fullerenes, that is, cages composed of sp^2 carbons, which contain only pentagons and hexagons. The pentagons must all be isolated following the isolated-pentagon rule (IPR). Experiment finds one single isomer, confirmed by ^{13}C nuclear magnetic resonance. Find the topology of this isomer.

a) Obtain the topologies of the two C_{76} isomers.

b) Calculate the energy eigenvalues of the two C_{76} isomers using Hückel theory.

c) Calculate the binding energy of each isomer.

d) Sketch the MO diagram (only some eigenvalues below and above the frontier orbitals) and fill the occupied energy levels following Hund's rule.

e) Find the most stable isomer. Discuss the result in terms of binding energy and occupation.

f) Explain why Hückel theory is suitable to make trend statements for fullerenes.

g) Bonus: The ^{13}C NMR pattern gives a line for each symmetrically distinct carbon atom with intensity proportional to the number of carbons of the same symmetry. Predict the number of lines and their intensity for both isomers.

Hints

A Hückel problem can be solved in the following way:

1. Set up the Hückel topology matrix.
2. Solve the eigenvalue problem. This is possible, for example, using computer algebra programs, but also using our distributed example program hueckel (see below).
3. Calculate (if necessary) the π electron energy, binding energy, and bond order.
4. Sketch the MO diagram and the occupation.
5. Sketch the MOs.

Fullerene topologies can, for example, be found in the literature [2]. Alternatively, they can be obtained by construction using the computer program CaGe (for a description, see Appendix), which contains a powerful fullerene generator.

7.5
Technical Details

For the Hückel calculations, the program hueckel, which is distributed together with this book, can be used. Since HMO theory is relatively simple, students with a background in programming may as well write their own code to solve this task. As well, they might inspect the source of the program (hueckel.f90), which is also distributed. The program hueckel needs an input file containing the topology of the system of interest in terms of a topology list (not matrix!), and the number of π electrons for calculation of the binding energy. For the Demonstration example, the input is

```
3   2
1   2   3   0
2   1   3   0
3   1   2   0
```

The first line gives the number of atoms and the number of p electrons. The following lines are numbered, so they carry the atom index. After the atom index, the program expects to read three other atomic indices that give its connectivity. If an atom is bound to one or more hydrogens (or other atoms, which are not relevant for Hückel theory), the connectivity index has to be set to 0. For the example given above, atom 1 is connected to atoms 2 and 3, atom 2 is connected to atoms 1 and 3, and atom 3 is connected to atoms 1 and 2. The order of the connectivity indices in the input file is irrelevant.

For simple systems, the topology input can be generated by hand. The fullerene structures are more complex. Therefore, the program CaGe (see the Appendix) or any other program that is able to obtain the topology of fullerenes can be used.

The input file should be given a self-explanatory name, for example, benzene.inp. On the Linux console, the Hückel calculation is started with the command

```
hueckel < benzene.inp > benzene.out
```

which writes the output into the file benzene.out. The output contains the eigenvalues and MO coefficients as well as the π electron energies and bond orders.

On the CD distributed with this book, we also supply an eigenvalue solver for general and special eigenvalue problems. You can use this as well in order to solve the Hückel problem, which is a special eigenvalue problem.

7.6
Review and Summary

In this chapter you have been introduced to Hückel molecular orbital theory in order to study conjugated carbon π systems. In particular, we have used HMO theory to calculate the molecular orbital energy levels, their eigenvector coefficients (MO coefficients), and the π and binding energies of planar sp^2 hybridized carbon-based systems. These have been used to discuss the stability of annulenes and fullerenes.

We have introduced the product ansatz in order to obtain single-particle equations instead of the full electronic Schrödinger equation. Moreover, the LCAO concept was used to represent the molecular orbitals in terms of atomic orbital contributions. The Hückel postulates further simplify the problem. These are the σ/π separation, the LCAO approximation, the neglect of orbital overlap resulting in orthonormal basis functions, and the approximation of the interaction potential present in the Hamilton integrals.

It is important to remember that only the topology (connectivity) of the molecule enters a Hückel calculation. Therefore, we have used the topology (Hückel) matrix (or, in practice, a topology list) and learned how to solve the Hückel problem by hand or using a computer. It is also vital to perform the re-substitution of x in the right way. The necessity of hard numbers for the parameters α and β has been discussed: it is usually not necessary to know the parameter α, since energies are given relative to it. Moreover, $\beta = -0.1$ Hartree is a good approximation to calculate energy differences.

Besides the energies, the MO coefficients are the most important result of a Hückel calculation. We have learned how to use them in order to plot the molecular orbitals. Remember that degenerate orbitals are special in two ways: first, you cannot tell which eigenvector belongs to which eigenvalue, and second, any orthonormalized linear combination of these eigenvectors satisfies the linear system of equations, that is, there is no single solution.

For further reading we refer the interested reader to the literature [3].

References

1 Atkins, P.W. and de Paula, J. (2006) *Physical Chemistry*, 4th edn, Oxford University Press.

2 Manolopoulos, D.E. and Fowler, P.W. (1995) *An Atlas of Fullerenes*, Oxford University Press.

3 Heilbronner, E. (1976) *The HMO Model and its Applications*, Vols I, II and III, John Wiley & Sons, Inc.

8
Hückel Molecular Orbital Theory:
Bond Order, Charge Order, and Molecular Orbitals

8.1
Aim

The HMO theory was introduced in the last chapter. Here, we will use it to introduce the concepts of bond order and charge order. The bond order concept is very helpful to locate double bonds in a planar sp^2 carbon structure. Moreover, we will discuss the molecular orbitals that are obtained through a Hückel calculation in more detail.

8.2
Theoretical Background

8.2.1
Bond Order

In the last chapter, Hückel molecular orbital theory [1, 2] was introduced and discussed with the focus on the energy eigenvalues and the MO coefficients. We have used quantities both to calculate the binding energy of molecules with conjugated π systems and to draw their molecular orbitals. Now we want to introduce two more quantities that can be calculated from the molecular orbitals of a Hückel calculation, which are determined by the eigenvectors. These quantities are the bond order and the charge order. The first one is well known from general chemistry and is crudely estimated as half of the number of bonding electrons minus half of the number of antibonding electrons in a bond. The second is the contribution of the Hückel system to the atomic charges in a molecule and thus a crude estimate of the atomic charge. Since HMO theory deals only with the π electrons of the system of interest, they should correctly be named π bond order and π charge order.

bond order

charge order

Computational Chemistry Workbook: Learning Through Examples
Thomas Heine, Jan-Ole Joswig, and Achim Gelessus
Copyright © 2009 WILEY-VCH Verlag GmbH & Co. KGaA, Weinheim
ISBN: 978-3-527-32442-2

π bond order   The bond order P_{kl} between the two atoms k and l is defined as

$$P_{kl} = \sum_{i=1}^{N} n_i c_k^{(i)} c_l^{(i)} \tag{8.1}$$

where $c_k^{(i)}$ and $c_l^{(i)}$ are the MO coefficients of orbital i at the atoms k and l, respectively, n_i is the occupation number of that orbital (being either 0, 1, or 2), and the orbital index i runs over all N (occupied and unoccupied) orbitals. Since for unoccupied orbitals $n_i = 0$, the sum runs effectively only over all occupied orbitals, which we can write as

$$P_{kl} = \sum_{i=1}^{occ} n_i c_k^{(i)} c_l^{(i)} \tag{8.2}$$

The bond order can be calculated between any two atoms, but it usually makes sense only for connected ones, because only these are directly bonded to each other. For normalized eigenvectors, the bond order can adopt values between 0 and 1. As a rough rule of thumb, we can state that π bond orders for bonded atoms larger than $\sim$0.7 indicate double bonds (the σ system yields another bonding pair of π electrons), and bond orders close to 0 indicate a single σ bond (the p system is either antibonding and neutralizes the σ bond, or there are no electrons in this bond at all, which usually implies that there is also no σ bond). This means that, from the bond order, we can determine the strength of the bond and thus also the π bond length. Note that the position of a double bond is hence the result of our calculation and is not implied a priori.

For the ethene molecule, for example, the coefficients of the lowest (and only) occupied MO are $c_1^{(1)} = c_2^{(1)} = 0.7071$, and as a result the bond order is

$$P_{12} = n_1 c_1^{(1)} c_2^{(1)} = 2 \times 0.7071^2 = 1.0000$$

The bond order thus describes the bonding situation in ethene correctly: a double bond.

8.2.2
Charge Order

π charge order   Since the MO coefficients give information about the negatively charged electron distribution in the molecule, we can use them to calculate the charge location, the so-called charge order. The charge order Q_k of atom k can be obtained as follows:

$$Q_k = \sum_{i=1}^{N} n_i (c_k^{(i)})^2 \tag{8.3}$$

Here, $c_k^{(i)}$ are the MO coefficients of orbital i at atom k, n_i is the occupation number of orbital i, and as above N is the number of carbon atoms and thus the number of eigenvalues. In contrast to the bond order, in which we used the MO coefficients of two atoms, we square all the coefficients of the occupied orbitals at one specific atom. Thus, the charge order shows how the π charge is distributed over the molecule. In the Demonstration that follows, this will be discussed in more detail.

8.3
Demonstration

8.3.1
Hückel Calculation of the Butadiene Molecule

As an example we will examine the butadiene molecule. It consists of four linearly connected sp² hybridized carbon atoms ($N = 4$) that each contribute one p orbital and thus one p electron to the π system. The total number of π electrons is four, and the Hückel calculation will result in four molecular orbitals. As before (cf. Chapter 7), we set up the topology **linear Hückel system** matrix, which has the typical structure of a linear Hückel system that is numbered through consecutively (remember that the numbering of the carbon atoms is essential for setting up the topology matrix and later drawing the molecular orbitals correctly):

$$\boldsymbol{H} = \begin{pmatrix} x & 1 & 0 & 0 \\ 1 & x & 1 & 0 \\ 0 & 1 & x & 1 \\ 0 & 0 & 1 & x \end{pmatrix} \quad \text{with} \quad x = \frac{\alpha - \varepsilon}{\beta} \tag{8.4}$$

Using the eigenvalue solver distributed with this book, the eigenvalues and eigenvectors listed in Table 8.1 are obtained.

In Table 8.1 we find four eigenvalues (numbered 1 to 4), which are the MO energies of butadiene (the lowest energy is given the index 1).

Table 8.1 Eigenvalues and eigenvectors resulting from a Hückel calculation of butadiene.

No.	Eigenvalue	Eigenvectors			
		Coefficient 1	Coefficient 2	Coefficient 3	Coefficient 4
4	−1.6180	0.3717	−0.6015	0.6015	−0.3717
3	−0.6180	−0.6015	0.3717	0.3717	−0.6015
2	0.6180	−0.6015	−0.3717	0.3717	0.6015
1	1.6180	0.3717	0.6015	0.6015	0.3717

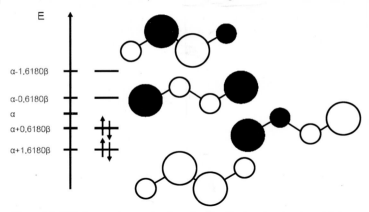

Figure 8.1 MO diagram (energy levels and occupation) and molecular orbitals of butadiene.

Corresponding to each eigenvalue, the MO eigenvectors (consisting of the four coefficients) are given.[1] The resulting MO diagram, including the orbital occupation, is shown in Figure 8.1. For a more detailed discussion, see Chapter 7. Note that the lowest energy orbital has no node and that the number of nodes increases with increasing energy. Remember that this representation implies that we are looking onto a π system from above. Thus, there is always a nodal plane in the plane of the paper.

Table 8.2 summarizes a number of properties of butadiene. The **frontier orbital** frontier orbitals, that is, the highest occupied molecular orbital (HOMO) and lowest unoccupied molecular orbital (LUMO), are $\varepsilon_2 = \alpha + 0.6180\beta$ and $\varepsilon_2 = \alpha - 0.6180\beta$ (note that β is negative), and thus the **HOMO–LUMO gap** HOMO–LUMO gap, the difference between HOMO and LUMO energies, is

$$\varepsilon_{gap} = |\varepsilon_2 - \varepsilon_3| = |(\alpha + 0.6180\beta) - (\alpha - 0.6180\beta)| = |1.2360\beta|.$$

Since we use $\beta = -0.1$ Hartree (cf. Chapter 7), we end up with a gap energy of $\varepsilon_{gap} = 0.1236$ Hartree $= 3.36$ eV. This is the lowest energy that is necessary to excite this molecule electronically. The wavelength λ of a photon can be calculated according to $\lambda = hc/E$, so that we need a wavelength of 369 nm to excite the butadiene molecule.

The binding energy was discussed in the previous chapter. Here, we will only have a look in more detail at the binding energy per atom, E_B/N. This is the average binding energy for each atom of the molecule, a

---

1) For example, the second eigenvalue is $\varepsilon_2 = \alpha + 0.6180\beta$, and the corresponding eigenvector is

$$c_2 = \begin{pmatrix} -0.6015 \\ 0.3717 \\ 0.3717 \\ -0.6015 \end{pmatrix}$$

Table 8.2 Different properties calculated for butadiene as defined throughout this chapter. Here, N is the number of C atoms, ε_{gap} the HOMO–LUMO gap, E_π the π electron energy, E_B the binding energy, P_{kl} the bond order, and Q_k the charge order. The energies have been calculated using $\beta = -0.1$ Hartree. The bond order has only been calculated for neighboring, that is, bonded, atoms.

N (no. of C atoms)		4
ε_{gap}		$1.2360|\beta|$
ε_{gap} (eV)		3.36
E_π		$4\alpha + 4.4720\beta$
E_B		4.4720β
E_B (eV)		-12.17
E_B/N (E_h)		-1.1180
E_B/N (eV)		-3.04
P_{kl}	P_{12}	0.8943
	P_{23}	0.4473
	P_{34}	0.8943
Q_k	Q_1	0.9999
	Q_2	0.9999
	Q_3	0.9999
	Q_4	0.9999

useful quantity for comparing the stability of structurally similar molecules of different size. For example, for a series of polyenes, the binding energy per atom may be calculated and its behavior as a function of polyene size may be analyzed. This will be done as the first problem.

The bond order for the bond between atoms 1 and 2 can be calculated using Equation 8.2:

$$
\begin{aligned}
P_{12} &= \sum_{i=1}^{4} n_i c_1^{(i)} c_2^{(i)} \\
&= n_1 c_1^{(1)} c_2^{(1)} + n_2 c_1^{(2)} c_2^{(2)} + n_3 c_1^{(3)} c_2^{(3)} + n_4 c_1^{(4)} c_2^{(4)} \\
&= 2 \times 0.3717 \times 0.6015 + 2 \times (-0.6015) \times (-0.3717) \\
&\quad + 0 \times (-0.6015) \times 0.3717 + 0 \times 0.3717 \times (-0.6015) \\
&= 2 \times 0.3717 \times 0.6015 + 2 \times (-0.6015) \times (-0.3717) \\
&= 0.8943 \qquad\qquad\qquad\qquad (8.5)
\end{aligned}
$$

As can be seen from the third line of Equation 8.5, only the occupied orbitals contribute to the bond order and also to the charge order. The latter for atom 3 is calculated as

$$Q_3 = \sum_{i=1}^{4} n_i (c_3^{(i)})^2$$

$$= n_1 (c_3^{(1)})^2 + n_2 (c_3^{(2)})^2 + n_3 (c_3^{(3)})^2 + n_4 (c_3^{(4)})^2$$

$$= 2 \times (0.6015)^2 + 2 \times (0.3717)^2 + 0 \times (0.3717)^2 + 0 \times (0.6015)^2$$

$$= 2 \times (0.6015)^2 + 2 \times (0.3717)^2$$

$$= 0.9999 \tag{8.6}$$

Analyzing the bond orders for all bonds and the charge orders for all atoms, we find that the π electrons are equally distributed over the butadiene molecule, that is, each atom has a π charge order of 1, i.e. a π charge of 0. The bond order shows an alternation of bond strength, that is, the bonds between atoms 1 and 2 and between atoms 3 and 4 have a much higher order than the central bond between atoms 2 and 3. In other words, the two terminal bonds have a double bond character and the central bond has a single bond character. This is what we would expect from chemical intuition. Note that we did not indicate the type of bond when performing the Hückel calculation. Only the bonding pattern enters the calculation. Thus, the result of the Hückel calculation is the location of the double bonds in a π system. Now, the molecule can be named buta-1,3-diene.

8.4
Problems

1. Stability of polyenes

a) Calculate the Hückel energy eigenvalues of the following polyenes: butadiene, hexatriene, octatetraene, and decapentaene.

b) Draw the eigenvalue spectra for each molecule and fill in the appropriate number of π electrons.

c) Calculate the difference between highest occupied and lowest unoccupied molecular orbitals (HOMO–LUMO gap) for each molecule. Draw the gap as function of chain length (number of carbon atoms).

d) Calculate the wavelength of the photons that are needed to excite the HOMO–LUMO transition in the different polyenes.

e) Calculate the π electron energy and the binding energy for all molecules according to Chapter 7.

f) Calculate the π bond order for the covalent bonds in all molecules.

g) Calculate the π bond order for selected non-bonded atoms.

h) Calculate the charge order of all atoms.

2. Bonding in the fulvene molecule

a) Use HMO theory to calculate the HOMO–LUMO gap, the π bond order, and the π charge of the fulvene molecule.

b) Sketch the occupied molecular orbitals.

c) Use a quantum chemical program to calculate MO energies, HOMO–LUMO gap, and orbital shapes within a better approximation of the Schrödinger equation.

d) The quantum chemical calculation yields many more orbitals than HMO theory. Find the subset of HMO-relevant orbitals among the MOs of fulvene that you have obtained by quantum chemical calculations. Compare their MO energies, their shape, and their relevance for the bonding of the molecule.

e) Which point group does fulvene belong to (cf. Chapter 2)?

Hints

1. The Hückel eigenvalue problem can be solved as described in Chapter 7. The supplied program `hueckel` calculates also the bond order and charge order. Nevertheless, you should first calculate the bond and charge orders by hand (at least for some small molecules) and then check your results using the program. A spreadsheet (Open Office Calc, Microsoft Excel) may be used as well.

2. The structure of fulvene is given in Figure 8.2. Bond lengths and angles are given in Ångstroms and degrees, respectively. For the HMO calculation, only the adjacency matrix is necessary. For the DFT calculation, in contrast, you have to set up a coordinate file from the given structural data.

8.5
Review and Summary

In this chapter we have introduced the concepts of π bond order and π charge order within Hückel molecular orbital theory. The bond order describes the double bond character of a bond between two atoms. It is calculated by summing the products of the MO coefficients of the two involved atoms. Values close to 1 indicate a double bond, values around 0 a single σ bond.

The charge order measures the charge location at the atoms. It is calculated from the MO coefficients of a single atom only by summing up the squared MO coefficients of the occupied orbitals. Both bond and

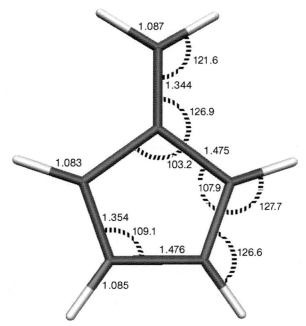

Figure 8.2 The D_{2h} structure of fulvene.

charge orders have been calculated for the butadiene system. The results have shown an alternation of double bonds. Their location in the molecule is thus the result of the bonding pattern.

The HOMO–LUMO gap energy has been calculated as well. Moreover, the wavelength necessary to excite an electron from the HOMO to the LUMO was obtained from the gap energy.

References

1 Atkins, P.W. and de Paula, J. (2006) *Physical Chemistry*, 4th edn, Oxford University Press.

2 Heilbronner, E. (1976) *The HMO Model and its Applications*, Vols I, II and III, John Wiley & Sons, Inc.

9
Geometry Optimization of a Diatomic Molecule

9.1
Aim

You will be introduced to methods of geometry optimization, in order to find a minimum energy structure of a molecule. We will first use the diatomic molecule carbon monoxide (CO), and approximate the interatomic interactions with a Morse potential. The geometry optimization will be performed first by hand and then using a more elaborate numerical optimization tool, provided by computational chemistry software.

geometry optimisation

9.2
Theoretical Background

The easiest way to understand the principles of molecular geometry optimizations is to perform a geometry optimization on a diatomic molecule step by step. After that, we can extend the theory easily to n-atomic molecules.

A diatomic molecule A–B has only one internal degree of freedom, the bond length R_{AB}. We will call it R in the following. To characterize the molecule, it is sufficient to know its stoichiometry and this single parameter R.

9.2.1
The Potential Energy Surface

Since the potential energy of a molecule depends on all the atomic coordinates, we usually deal with a multi-dimensional function. For a diatomic molecule, such as those we will work within in this chapter, the problem is relatively easy: just the bond length determines the potential energy. Here, we discuss the potential V acting on the nuclei

Computational Chemistry Workbook: Learning Through Examples
Thomas Heine, Jan-Ole Joswig, and Achim Gelessus
Copyright © 2009 WILEY-VCH Verlag GmbH & Co. KGaA, Weinheim
ISBN: 978-3-527-32442-2

of the molecule. Following the Born-Oppenheimer approximation (see Chapter 15), we treat the system of nuclei classically using Newtonian mechanisms [4]. The potential is equivalent to the potential energy, and we use synonymously $V(R) = E_{pot}(R)$. For higher- dimensional func-

potential energy (hyper) surface

tions $V = V(R_1, R_2, \ldots, R_N) = E_{pot} = E_{pot}(R_1, R_2, \ldots, R_N)$, we call the function a potential energy (hyper)surface (PES). Note that in the literature you will often find the term total energy instead. Total energy includes the expectation value of the electronic Hamiltonian, the electronic energy, and the nucleus–nucleus repulsion, but does assume that nuclei are at rest. If the nuclei are moving, as in a molecular dynamics simulation, which we will discuss in Chapter 15, the total energy also contains the kinetic energy of the nuclei. As we will keep this in mind, we will speak here about potential energy and PES.

Depending on the description of the interatomic interactions, this potential energy surface may have various maxima, minima, and saddle points. For the purpose of a geometry optimization, we usually seek a potential energy minimum, within a certain range possibly the

global minimum

lowest one, the so-called global minimum of the PES. Local PES minima are points on the PES (i.e., special atomic arrangements), that do not have points of lower energy in their neighborhood (Figure 9.2). Let us discuss the PES of a simple molecule, difluoroethene. Obviously, we can sketch three different structures for this molecule, as shown in Figure 9.1.

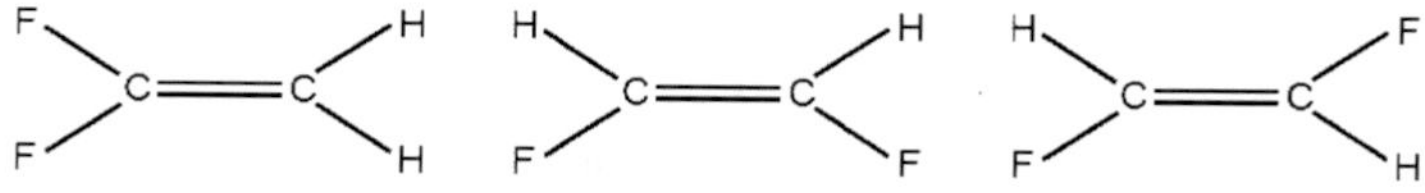

Figure 9.1 Three structural isomers of difluoroethene (left to right): 1,1-$C_2H_2F_2$, *cis*-$C_2H_2F_2$, and *trans*-$C_2H_2F_2$.

isomer

Each of these isomers has a corresponding potential energy. All three structures correspond, thus, to local minima on the PES, but only one of them corresponds to the global minimum, since it has a lower energy than the other two or any other, even chemically meaningless, arrangement. To know which of the structures are stable, we need to compare the energy of the three minima. This will be elaborated in Chapter 14. Briefly, a local minimum on the PES is stable if it has a reasonably low energy, and if it is separated from close, lower local minima by high energy barriers (see Figure 9.2).

For a diatomic molecule, the PES is rather simple and approximately given by the Morse potential (see Equation 3.1 and Figure 3.1). This potential has only one minimum, the global minimum, which gives the equilibrium structure of the molecule.

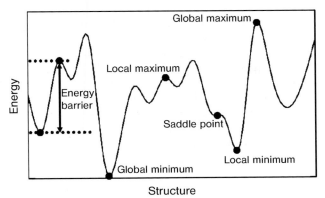

Figure 9.2 Schematic (two-dimensional) view of a potential energy surface.

9.2.2
Forces in a Diatomic Molecule

Finding the equilibrium structure of a diatomic molecule is almost the same exercise as finding the valley between two mountains: the landscape is the potential energy surface. We just need to follow the gravitational force, which brings us downhill. Once there is no possibility of further decrease (in any direction), we have found the minimum.

diatomic molecule

If we translate this approach to a diatomic molecule, we need to find the position where the potential energy has its minimum. Consequently, the derivative of the potential with respect to R equals zero. We know from Chapter 3 that the intermolecular force is the negative of the derivative of the potential energy with respect to the interatomic distance $F(R) = -\partial V(R)/\partial R$. At the minimum, the force disappears. We can transform the interatomic force to the force acting on both atoms A and B, which follows the same steps as for the derivation of Equation 3.14, and obtain $F_A = F_B = 0$. Now, we use the fact that a net force always points to the direction where the potential energy is lower. Let us further generalize that we are working in three-dimensional space, so the forces are three-component vectors and at the minimum $\mathbf{F}_A = \mathbf{F}_B = 0$ holds.

What would happen for a molecule with more than two atoms? In the equilibrium structure, all the forces on the N atoms must vanish as well, as otherwise the atoms would start to move, the potential energy would get lower, and hence the molecule would not have been at its equilibrium structure. We can again find the equilibrium structure by following the forces until we reach the position of the PES, where the equilibrium criterion $\mathbf{F}_1 = \mathbf{F}_2 = \ldots = \mathbf{F}_N = 0$ holds.

The force vector on the atom with index I is computed, independently of the number of atoms in the molecule, directly as the derivative of the potential energy surface with respect to the position vector $\boldsymbol{R}_I$ of atom I,

$$\boldsymbol{R}_I = \begin{pmatrix} X_I \\ Y_I \\ Z_I \end{pmatrix}$$

that is,

$$\boldsymbol{F}_I = -\frac{\partial V}{\partial \boldsymbol{R}_I} = \begin{pmatrix} -\partial V/\partial X_I \\ -\partial V/\partial Y_I \\ -\partial V/\partial Z_I \end{pmatrix} = \begin{pmatrix} F_{I,X} \\ F_{I,Y} \\ F_{I,Z} \end{pmatrix} \tag{9.1}$$

If the molecule is in equilibrium, $\boldsymbol{F}_I = 0$ for all atoms I.

For a diatomic molecule, the second derivative of the potential with respect to the interatomic distance must be positive, that is,

$$\frac{\partial \boldsymbol{F}}{\partial \boldsymbol{R}} = -\frac{\partial^2 V}{\partial \boldsymbol{R}^2} > 0 \tag{9.2}$$

Equation 9.2 is a strict condition to characterize a local minimum – a negative value would characterize a maximum of the potential – and this is known as the transition state. In the case of more than two atoms, the minimum condition turns out to be more complex. In this case, the Hesse matrix (also called Hessian or Hessian matrix) needs to be calculated and diagonalized. The Hessian is a $3N \times 3N$ matrix, defined as

Hessian matrix

$$\boldsymbol{H} = \left(\frac{\partial^2 V(\boldsymbol{R}_1, \boldsymbol{R}_2, \ldots, \boldsymbol{R}_N)}{\partial \boldsymbol{R}_I \, \partial \boldsymbol{R}_J} \right) \tag{9.3}$$

This matrix needs to be computed and diagonalized. 6 (5 for linear molecules) eigenvalues are zero (or very small) and correspond to rotations and translations. As these do not affect the molecular structure, we can omit them. If all remaining eigenvalues are positive, the molecule is in a minimum of the PES; if one or more eigenvalues are negative, the molecule is in a transition state. The eigenvalues of the Hessian are directly related to the vibrational frequencies of the molecule. We will therefore postpone the discussion on the Hessian to Chapter 11.

9.2.3
The Steepest Descent Method

steepest descent method The steepest descent method is certainly the simplest method to find an atomic arrangement for which all the net forces on the atoms vanish. To find a local minimum, a steepest descent uses the gradient of some function, in our case of the potential energy V, so it is directly related to the net forces on the atoms (see Equation 9.1). This is comparable to descending from a mountain: following gravity, which valley you reach depends on which side of the mountain you start descending. Thus, the starting point (the starting geometry) is most important for the result.

In the numerical implementation of the steepest descent method, one follows the forces on the atoms in an iterative procedure. A new **iterative procedure** geometry is calculated by making a step along the direction of the force vectors:

$$\boldsymbol{R}_I^{(j+1)} = \boldsymbol{R}_I^{(j)} + \boldsymbol{S}_I^{(j)}, \quad \boldsymbol{S}_I^{(j)} = \gamma \boldsymbol{F}_I^{(j)} \qquad (9.4)$$

Here, $\boldsymbol{R}_I^{(j)}$ is the position vector of the Ith atom in the jth iteration step; consequently $\boldsymbol{F}_I^{(j)}$ is the force acting on it. The steps $\boldsymbol{S}_I^{(j)}$ are added to the present position vectors and are proportional to the forces. The multiplier γ ($\gamma > 0$) scales the forces to have reasonable steps $\boldsymbol{S}_I^{(j)}$, and hence a reasonable change of geometry. As the method is supposed to follow the steepest descent, γ has to be chosen small enough to avoid too large geometry changes, otherwise spurious effects may occur. In the worst case, for very large forces as are sometimes present for initial structures, the geometry might break if too large values of γ are chosen – so it is always a good strategy to define a maximum step size to avoid problems. The iteration may be stopped if all the forces are below a certain threshold, or if the potential energy does not change within a tolerance between two successive steps.

The potential energy is the ideal quantity to monitor during the steepest descent algorithm. If the energy increases in a new iteration, then the minimum must be between the last two geometries. If γ is reasonably small, it means that the structure is very close to a minimum. To find the equilibrium structure, one goes back to the previous structure and further decreases γ to decrease the step size.

9.2.4
Hessian-Based Optimizers and the Newton–Raphson Method

The drawback of the steepest descent method is its inefficiency, in particular if one comes close to the minimum or if the PES is very flat. A possibility to reduce the number of iterations is to use not only the gradient but also the curvature of the PES at the present point. We will illustrate this Hessian-based Newton–Raphson method for the **Hessian-based Newton–Raphson method** one-dimensional problem.

Equation 9.4 is a linear extrapolation of the position vector. For large gradients, large steps are taken; for small gradients, the steps are rather small. If, for example, a flat minimum is found, then the number of iterations for Equation 9.4 will be very large. We can extrapolate more precisely if we also take advantage of the curvature of the potential, its second derivative. The second derivative is then used to have a variable value for γ in Equation 9.4, which is updated as

$$R_I^{(j+1)} = R_I^{(j)} + S_I^{(j)}, \quad S_I^{(j)} = H^{-1}F_I^{(j)} \tag{9.5}$$

using the Hesse matrix H as defined in Equation 9.3. Strict application of Equation 9.5 will lead to numerical problems. The Hessian is not invertible, as it includes the external degrees of freedom (translation and rotation of the whole molecule, which do not change the potential energy). Transformation of the Cartesian coordinates to internal co-ordinates solves this problem, but makes the mathematical handling more tedious, as bond angles now have to be treated. Internal redundant coordinates are a modern alternative that we recommend to interested students for further reading [1].

In one dimension, there is only one value $R^{(j)}$ to be iterated, and the Hesse matrix reduces to a single number. We will continue to discuss this simplified case in this chapter. Equation 9.5 is nothing else than expecting that the harmonic approximation holds well for the potential at the present position. For such a truly harmonic function, Equation 9.5 approaches the minimum in a single step. It is obvious that Equation 9.5 will be an excellent way to find a minimum if we are already in its close vicinity. For geometries far from a local minimum, or for a very anharmonic PES, application of Equation 9.5 may fail.

In the demonstration in this chapter, optimizations following the schemes of Equation 9.4 (steepest descent optimizer) and Equation 9.5 (Hessian-based Newton–Raphson optimizer) are compared.

In practice, direct application of Equation 9.5 to larger molecules is not reasonable, as the calculation of the Hessian would require too much computer time. Therefore, one starts with a steepest descent step, and updates the Hessian using the finite differences of a history of succeeding geometry optimization steps. Various update schemes are known in the literature, the most prominent one being that proposed independently by Broyden, Fletcher, Goldfarb and Shanno in 1970, the BFGS method [2]. The method also depends crucially on the employed coordinate system. Hessian-based optimization methods are presently the standard methods in chemistry. However, for large systems they are inadequate, as handling of the huge Hessian becomes problematic, and more robust algorithms, for example based on the conjugate gradient method, are more efficient. Another drawback of these methods is that, if the Hessian is updated during successive geometry steps and if the potential energy surface is not well approxi-mated using a second-order Taylor series expansion, then the Hessian is badly approximated, and Equation 9.5 might lead to very poor steps, and hence poor convergence. In such cases, the optimization must be restarted.

9.2.5
Application to the Morse Potential

For demonstration purposes, we will apply the two algorithms of Equations 9.4 and 9.5 to the Morse potential, which we know already from Chapters 3 and 4. It is a quite good approximation of the PES of diatomic molecules, and we know its analytic form from Equation 3.1:

Morse potential

$$V(R) = D_e\{[1 - e^{-\alpha(R-R_e)}]^2 - 1\}$$

Here, we have already transformed the Cartesian coordinates of the molecule to internal coordinates, leaving the intermolecular distance as the single variable. Transformation back to Cartesian space is equivalent to the derivation of Equation 3.14. All optimization approaches can be applied to systems in internal coordinates, and the particularly simple situation of a single variable is convenient for both methods discussed here.

Note that Equations 9.4 and 9.5 simplify significantly as we only have to treat a one-dimensional problem, since the molecule is *by definition* aligned with the R axis. It is convenient to calculate the first and second derivatives of Equation 3.1 to apply them in the iteration schemes. We know them from Chapter 3 (Equation 3.15):

$$\frac{\partial V(R)}{\partial R} = 2D_e\alpha[e^{-\alpha(R-R_e)} - e^{-2\alpha(R-R_e)}] = -F(R),$$

$$\frac{\partial^2 V(R)}{\partial R^2} = 2D_e\alpha^2[2e^{-2\alpha(R-R_e)} - e^{-\alpha(R-R_e)}]$$

9.2.6
Global Versus Local Geometry Optimization

Using gradient-based optimization methods, the result of a geometry optimization depends strongly on the starting configuration of the atomic coordinates. This means that, for a given stoichiometry, there is often more than only one energy minimum, that is, there are several stable isomers. After an optimization of all those isomers, it is possible to decide which of them has the lowest total energy by comparing their potential energies. The minima obtained from a geometry optimization are therefore called local minima on the potential energy surface; the lowest energy minimum is called the global minimum. An example is the hydrogen cyanide (HCN) molecule, of which two isomers exist: $H-C\equiv N$ and $C=N-H$. Depending on the starting configuration of the atoms, the geometry optimization might reach one or the other structure.

geometry optimization

In chemistry, the number of possible isomers is strongly reduced by the fact that not all atomic arrangements are chemically possible, for example, there will be no C−H−N isomer. In the physics of clusters and nanostructures, the number of local minima might be much larger. In Chapter 7 you have already compared the total energy obtained from a Hückel calculation of different fullerene isomers. Take the time and try to construct all the possible isomers of a larger fullerene, for example C_{90}, using the CaGe program. Then you will understand that global geometry optimization [3] is beyond the scope of this book.

9.3
Demonstration

We will optimize the carbon monoxide (CO) molecule using the steepest descent method and the Hessian-based Newton–Raphson method. In both cases, we start the minimum search at (i) $R^{(0)} = 80$ pm and (ii) $R^{(0)} = 150$ pm, so that we approach the equilibrium distance from below and above. We have chosen pm, 100 pm $= 1$ Å, as length unit, as this is the most convenient for notation in this particular experiment. We use the Morse potential of Equation 3.1 with the parameters taken from Table 3.1, namely $D_e = 1072.8$ kJ mol^{-1}, $\alpha = 2.312 \times 10^{-2}$ pm^{-1}, and $R_e = 113$ pm, for which we have converted all values except the energy into SI units. Note that the result (R_e) is already given in the problem, but as an exercise we will seek it here numerically, as this makes you familiar with how computational chemistry software – in principle – works. In a geometry optimization, we want to determine the potential energy of the optimized molecule in order to be able to compare the potential energies of different isomers. Again, the result D_e is given within the parameters of the Morse potential. In this chapter we deliberately choose some problematic cases to illustrate the importance of controlling the numerical procedure in a geometry optimization. For convenience, it is advisable to use a spreadsheet (see Technical Details) for the forthcoming calculations.

9.3.1
Optimizing CO With the Steepest Descent Method

In the one-dimensional case, Equation 9.4 simplifies to $R^{(j+1)} = R^{(j)} + S^{(j)}$ with $S^{(j)} = \gamma F^{(j)}$. From the starting distance $R^{(0)} = 80$ pm, we can calculate the potential using Equation 3.1: $V(R^{(0)}) = V(80\,\text{pm}) = 333$ kJ mol^{-1}. Note that this energy is positive, which means that we are in the repulsive part of the potential, and a strong force can be expected. Following Equation 3.15, the force is $F^{(0)} = F(R^{(0)}) = F(0.8 \times 10^{-10}\,\text{m}) = 122$ kJ mol^{-1} pm^{-1}. This means that changing the bond length by 1 pm results in a predicted change in energy of $122\,\mathrm{kJ\,mol^{-1}}$, obviously a quite large value compared to the value of the potential which indicates that the potential varies strongly at $R^{(0)}$. The unit is intuitively clear for a chemist, but is somewhat unconventional. It can, of course, easily be converted into the SI unit, $F^{(0)} = 2.0 \times 10^{-7}\,\mathrm{N}$. The force has, as expected, a positive sign, pointing towards larger values of R. We now have to choose a γ so that the step $S^{(0)}$ has a reasonable order of magnitude, and it changes R only by a small distance. If no further information about the system is known, it is safe to use a small value, as usually – but not always – forces will reduce during the optimization. In this case, a value of $\gamma = 0.1\,\mathrm{mol\,pm^2\,kJ^{-1}}$ gives a reasonable initial step $S^{(0)} = \gamma F^{(0)} = (0.1\,\mathrm{mol\,pm^2\,kJ^{-1}}) \times (122\,\mathrm{kJ\,mol^{-1}\,pm^{-1}}) = 12.2\,\mathrm{pm}$. The resulting step size is in the order of a tenth of the bond length, which is an acceptable value. Moreover, the step size is positive, so the bond length increases in the first step by $0.122\,\text{Å}$. The first iteration gives $R^{(1)} = R^{(0)} + S^{(0)} = 80\ \mathrm{pm} + 12.2\,\mathrm{pm} = 92.2\,\mathrm{pm}$, which is already considerably closer to the result than the starting point. We proceed now with the iterative procedure to find the equilibrium structure. The results are given in Table 9.1

Obviously, the performance of this method is not acceptable. We have to restrict the multiplier of Equation 9.4 to a very small value to avoid unphysical initial steps ($\gamma = 0.1\,\mathrm{mol\,pm^2\,kJ^{-1}}$), which results in very poor performance for positions closer to the equilibrium structure, that is, we have to perform a large number of iterations. An alternative is to have a larger multiplier, but to restrict the maximum step, by modification of Equation 9.4 to

$$R^{(j+1)} = R^{(j)} + S^{(j)}, \quad S^{(j)} = \frac{F^{(j)}}{|F^{(j)}|}\,\min\!\left(\Delta, \gamma|F^{(j)}|\right) \qquad (9.6)$$

Table 9.1 Steepest descent geometry optimization for CO, starting at $R^{(0)} = 0.8\,\text{Å} = 80\,\mathrm{pm}$, and a value of $\gamma = 0.1\,\mathrm{mol\,pm^2\,kJ^{-1}}$. See text for further details.

Iteration j	$R^{(j)}$ (pm)	$V(R^{(j)})$ (kJ mol⁻¹)	$F(R^{(j)})$ (kJ mol⁻¹ pm⁻¹)	$S^{(j)}$ (pm)
0	80	332.7	121.8	12.2
1	92.2	−662.6	49.6	4.96
2	97.1	−862.4	31.70	3.17
3	100.3	−948.1	22.67	1.17
10	109.0	−1062.7	5.276	0.528
20	111.94	−1071.14	1.261	0.144
50	112.974	−1072.7996	0.030	0.003
100	112.99994	−1072.799999998	0.0000683	0.00000683

Table 9.2 Steepest descent geometry optimization for CO, starting at $R^{(0)} = 80$ pm, and a value of $\gamma = 1$ mol pm^2 kJ^{-1}. The step size has been limited to $\Delta = 10$ pm.

Iteration j	$R^{(j)}$ (pm)	$V(R^{(j)})$ (kJ mol^{-1})	$F(R^{(j)})$ (kJ mol^{-1} pm^{-1})	$S^{(j)}$ (pm)
0	80.0	333.72	121.8	10[a]
1	90.0	−544.23	59.3	10[a]
2	100.0	−940.92	23.5	10[a]
3	110.0	−1067.27	3.82	3.82
4	113.82	−1072.42	−0.913	0.913
5	112.91	−1072.795	0.108	0.108
6	113.014	−1072.79989	−0.0162	−0.0162
7	112.9979	−1072.799998	0.00238	0.00238
8	113.0003	−1072.79999995	−0.000349	−0.000349
9	112.999955	−1072.7999999989	0.00000754	0.0000754

[a]The step size has been limited to a maximum of 10 pm.

Equation 9.6 is given only for one dimension, but it is straightforward to extend it to a whole molecule. The iterative procedure as given in Equation 9.6 restricts the geometry step to a maximum value Δ (a reasonable value for minimum searches is $\Delta = 10$ pm) and always points into the direction of the force. As we avoid numerical problems of too large step sizes this way, we can safely increase the multiplier by a factor 10 to $\gamma = 1$ mol pm^2 kJ^{-1}. The modified algorithm has been applied and the results are given in Table 9.2.

Obviously, we approach the minimum much faster. The procedure could even be improved further by introducing a variable γ that is linked to the magnitude of the force. However, the method is still far from being perfect. For example, if oscillations around the minimum appear, the iterations may not converge. Before we proceed to the Hessian-based Newton–Raphson method we will approach the minimum from a large distance, with $R^{(0)} = 150$ pm. The first iterations are given in Table 9.3

9.3.2
Optimizing CO Using a Hessian-Based Newton–Raphson Optimizer

The Hessian matrix of the CO molecule, which is uniquely defined by the bond length, is

$$\boldsymbol{H} = (H_{11}) = \left(\frac{\partial^2 V(R)}{\partial R^2} \right)$$

that is, it is the second derivative of the Morse potential with respect to the bond length. We know this derivative from Equation 3.15.

Table 9.3 Steepest descent geometry optimization for CO, starting at $R^{(0)} = 150$ pm, and a value of $\gamma = 1$ mol pm^2 kJ^{-1}. The step size has been limited to 10 pm.

Iteration j	$R^{(j)}$ (pm)	$V(R^{(j)})$ (kJ mol^{-1})	$F(R^{(j)})$ (kJ mol^{-1} pm^{-1})	$S^{(j)}$ (pm)
0	150.0	−718.224	−12.123	−10^a
1	140.0	−841.500	−12.338	−10^a
2	130.0	−959.487	−10.882	−10^a
3	120.0	−1048.847	−6.305	−6.305
4	113.695	−1072.527	−0.77841	−0.77841
5	112.917	−1072.7960	0.09563	0.09563
6	113.012	−1072.79991	−0.014317	−0.014317
7	112.9982	−1072.799998	0.0020971	0.0020971
8	113.0003	−1072.79999996	−0.0003082	−0.0003082
9	112.99996	−1072.7999999991	0.00004527	0.00004527

^{a}The step size has been limited to a maximum of 10 pm.

The Hessian matrix in this case is a 1×1 matrix containing just one element H_{11}. The inverse of the Hessian matrix, which we need according to Equation 9.5, is then the reciprocal of H_{11}. Again, we will start the iterations with a too short bond length, $R^{(0)} = 80$ pm. Table 9.4 lists the first iteration steps. First, the potential energy $V^{(j)}$ is calculated (not mandatory, but helpful). Second, the force $F^{(j)}$ is calculated. Both steps are similar to the steepest descent method. In fact, they do not belong to the actual optimization steps. As a third step, the second derivative (Equation 3.15) is calculated at point $R^{(j)}$, which is in our case the one single Hessian matrix element H_{11}. Finally, the step is obtained by the relation $F^{(j)}/H_{11}^{(j)}$. Note that we do not need to scale the forces as in the steepest descent method. The interatomic distance of the next iteration is calculated by adding the step to the old distance.

Table 9.4 Hessian-based Newton–Raphson geometry optimization for CO, starting at $R^{(0)} = 80$ pm.

Iteration j	$R^{(j)}$ (pm)	$V(R^{(j)})$ (kJ mol^{-1})	$F(R^{(j)})$ (kJ mol^{-1} pm^{-1})	$H_{11}^{(j)}$ (kJ mol^{-1} pm^{-2})	$S^{(j)}$ (pm)
0	80	332.72	121.77	8.090	15.05
1	95.05	−789.0095	38.636	3.523	10.97
2	106.02	−1039.8696	10.2138	1.820	5.612
3	111.629	−1071.68717	1.6491	1.260	1.3088
4	112.9377	−1072.797768	0.07165	1.152	0.06220
5	112.999866	−1072.79999999	0.0001543	1.146908	0.000134
6	112.9999999994	−1072.8	0.00000000072	1.146897	0.00000000063

The result is the same as for the steepest descent method: the interatomic distance and the corresponding potential energy are the same. The method is, however, significantly faster in convergence, in particular when demanding a high accuracy. It would have been reasonable to stop the iteration after step 5, because continuing did not alter the result significantly. However, the value for $F^{(6)}$ shows that we can get the forces easily to practically zero.

A fast method also has its pitfalls, which can be seen if we start from, $R^{(0)} = 150$ pm. If we calculate H_{11}, we obtain a negative value, which means that Equation 9.5 would reverse the direction of the step, and wrong results would be produced. A negative value for H_{11} is clearly far away from the harmonic area of the potential, which is what is assumed in Equation 9.5. In such cases, which are easily detected, one would need to perform a couple of steepest descent steps. After some iterations, the sign of H_{11} changes. Again, we obtain a numerical problem, as the prediction of the step is in the order of 100 pm. This tells us that it is always a good idea to limit the step size to maximum. Once we respect these conditions, use a value for γ as before for the steepest descent method, and restrict the maximum step size to 10 pm, the molecule finds its equilibrium geometry quickly (see Table 9.5). The modified Equation 9.5 for a single variable reads as

$$R^{(j+1)} = R^{(j)} + S^{(j)}, \quad S^{(j)} = \frac{F^{(j)}}{|F^{(j)}|} \min\left(\Delta, \max\left([H_{11}^{(j)}]^{-1}, \gamma\right) \times |F^{(j)}|\right)$$

$$(9.7)$$

Table 9.5 Hessian-based Newton–Raphson geometry optimization for CO, starting at $R^{(0)} = 150$ pm. The step size has been limited to 10 pm. For negative values for the Hessian, the step was replaced by a steepest descent step with a value of $\gamma = 1$ mol pm^2 kJ^{-1}.

Iteration j	$R^{(j)}$ (pm)	$V(R^{(j)})$ (kJ mol^{-1})	$F(R^{(j)})$ (kJ mol^{-1} pm^{-1})	$H_{11}^{(j)}$ (kJ mol^{-1} pm^{-2})	$S^{(j)}$ in pm
0	150	−718.224	−12.123	−0.073	−10[a]
1	140	−841.500	−12.338	0.0438	−10[a]
2	130	−959.487	−10.882	0.27096	−10[a]
3	120	−1048.847	−6.3147	0.683993	−9.218
4	110.782	−1069.831	2.7468	1.33425	2.0588
5	112.8412	−1072.7855	0.183108	1.15958	0.15791
6	112.99913	−1072.7999996	0.000997	1.146966	0.0008697
7	112.99999997	−1072.8	0.000000030	1.146897	0.000000026

[a]The step size has been limited to a maximum of 10 pm.

9.4
Problem

1. Set up your environment

Program a spreadsheet or computer program that is reproducing the numbers in Tables 9.1 to 9.5. This program can then be used for the following tasks. All geometry optimizations have to be carried out with a tolerance of 10^{-3} pm for the bond length, that is, the bond lengths of two successive iterations differ by less than 10^{-3} pm.

2. Compare the importance of the starting point

Optimize the geometry of CO starting from 90, 100, 110, 120, and 130 pm. How many iteration steps are needed in the modified steepest descent (Equation 9.6) and Hessian-based Newton–Raphson optimizers (Equation 9.7)?

3. Optimize the geometry for H_2, HI, and I_2

a) Optimize the geometries of the three molecules hydrogen (H_2), hydrogen iodide (HI), and iodine (I_2) with both algorithms, starting with a bond length of $0.8R_e$ and $1.2R_e$, respectively.

b) Optimize the geometry of the three molecules using modern computational chemistry software. Note that the results will differ somewhat as a different potential energy surface is used.

4. Familiarize yourself with the performance of modern geometry optimizers

a) Sketch starting structures of the following molecules using a molecular editor: water, ethane, furane, o- and p-fluorobenzene. Compare your results to available experimental data.

b) Optimize the geometry using modern computational chemistry software.

9.5
Technical Details

1. In any environment you can use a spreadsheet, for example, Open Office Calc and Microsoft Excel.

2. For problem 3 you can use `Molden` to sketch the molecules. `Molden` is also an excellent tool to analyze the geometry optimization, since it can visualize the energy and the maximum gradient at each iteration step. The change of structure can be shown in a movie clip (press the **Geom. Conv.** button; also, you may use the **Point Selection** part to see the structural change during the optimization).

3. If you use deMon for performing the geometry optimization, use an adopted input from the following example for CO:

```
TITLE CO GEOMETRY OPTIMISATION
VXCTYPE AUXIS PBE
BASIS (DZVP-GGA)
AUXIS (GEN-A2)
OPTIMISATION
VISUALISATION MOLDEN
GEOMETRY Z-MATRIX
C
O 1 RCO
VARIABLES
RCO 0.8
END
```

9.6
Review and Summary

In this chapter we have performed geometry optimizations, in order to find the minimum energy structure of a molecule. We started with the diatomic molecule, carbon monoxide, as a demonstration example, and have approximated the interatomic interactions with a Morse potential. Moreover, we have optimized larger molecules using more advanced optimization tools and a potential energy surface computed on the grounds of density-functional theory.

In the numerical procedure, the forces acting on the individual atoms have to be determined. The forces have been calculated as the negative of the first derivative of the potential energy with respect to the atomic positions. In the three-dimensional structure of molecules, the forces are three-dimensional vectors, and the geometry steps are vectors with contributions in X, Y, and Z directions. In order to reach a local minimum of the PES, the atomic positions are changed iteratively, and at each step the forces are recalculated until they are below a defined threshold.

We have reviewed two methods to find a local minimum of the PES. The steepest descent method is a very simple and robust method that searches for the minimum along the steepest descent. However, it has the drawback of efficiency, since many steps are needed to find a minimum. This behavior is improved in Hessian-based optimization tools, where the prediction of geometry steps is significantly better, and many fewer iterations are necessary to find a minimum. The latter methods have the drawback that the calculation of the Hessian is very time consuming.

Alternatively, approximate Hessians can be built, either by lower-level methods, or by updating the Hessian from earlier iteration steps. Both approaches have the risk of a poor approximation of a Hessian. In practice, if a geometry optimization takes too many iteration steps, the update should be restarted.

References

1 Eckert, F., Pulay, P. and Werner, H.-J. (1997) *Ab initio* geometry optimization for large molecules. *Journal of Computational Chemistry*, **18**, 1473–1483.

2 Avriel, M. (2003) *Nonlinear Programming: Analysis and Methods*, Dover Publications.

3 Wales, D.J. (2003) *Energy Landscapes*, Cambridge University Press.

4 Born, M. and Oppenheimer, J.R. (1927) *Zur Quantentheorie der Molekeln*. Annalen der Physik **389**, 457–484.

10
The Electron Spin

10.1
Aim

Electrons are fermions and have an electron spin. In this experiment we study molecules that have a net spin due to unequal numbers of spin-up and spin-down electrons. We will understand why such compounds, radicals, exist, and we will learn to determine the energetically favored distribution of spin-up and spin-down electrons in a molecule.

10.2
Theoretical Background

10.2.1
The Electron Spin

The Pauli exclusion principle [1] states that no two identical fermions may occupy the same quantum state simultaneously. Electrons, being fermions with electron spin $\sigma = \pm \frac{1}{2}\hbar$, are subject to the Pauli principle. In most cases the consequence is that all electronic states are occupied by two electrons, one with $\sigma = +\frac{1}{2}\hbar$ and one with $\sigma = -\frac{1}{2}\hbar$, so that all the electrons are paired, and hence the total spin of all the electrons of the molecule sums up to zero. Those molecules are called diamagnetic, and all closed-shell molecules have this property.

    There are several conditions that lead to situations where the number of spin-up and spin-down electrons is not equal. The most obvious case is a molecule that contains an odd number of electrons (e.g., NO), but there are also cases where symmetry creates degenerate electron levels and there are not enough electrons to fill them, like for example in O_2. Such molecules are paramagnetic and are called radicals and open-shell molecules. If we go to the heavier elements in the periodic table, in particular to the f elements, the atoms may have a partially occupied shell that contains unpaired electrons but does not participate in the

Pauli exclusion principle

electron spin
fermion

diamagnetic
closed-shell molecule

paramagnetic
open-shell molecules

Computational Chemistry Workbook: Learning Through Examples
Thomas Heine, Jan-Ole Joswig, and Achim Gelessus
Copyright © 2009 WILEY-VCH Verlag GmbH & Co. KGaA, Weinheim
ISBN: 978-3-527-32442-2

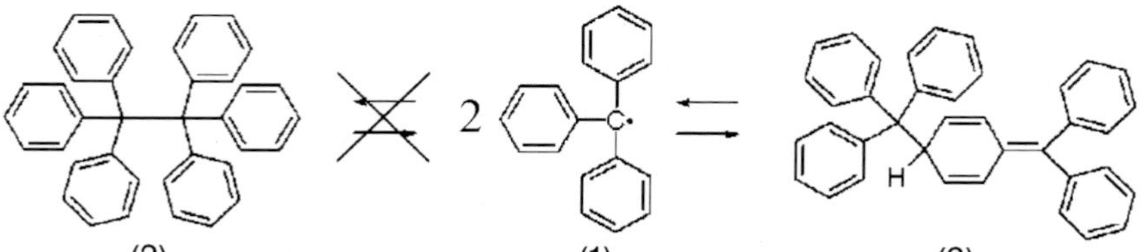

Figure 10.1 The triphenylmethyl radical (1) cannot form a symmetric dimer (2) for steric reasons. The asymmetric dimer (3) is close in energy to two monomer radicals.

bonding. The electrons in these shells remain unpaired when molecules are formed, which are consequently paramagnetic (e.g., myoglobin). In this experiment, however, we will concentrate on situations where unpaired electrons appear in the frontier orbitals, as is the case for radicals.

Most radicals are chemically reactive and therefore rapidly transformed into more stable closed-shell molecules. Still, if they appear in low concentrations, they may have considerable lifetimes. This fact is used for various chemical reaction mechanisms, for example in polymerizations. The situation may be different if the frontier orbitals are delocalized, for example in aromatic rings. Here, radicals can be quite stable. A prominent example is the triphenylmethyl radical, which is in equilibrium with its dimer at room temperature in benzene solution (Figure 10.1).

10.2.2
The Multiplicity

spin angular momentum To simplify the notation, we introduce the spin angular momentum of a molecule, S. This gives the net spin in units of $\hbar$. An electron contributes
multiplicity to S with $\pm\frac{1}{2}$. The multiplicity $M = 2S + 1$ is a related quantity. As it is always an integer, it is more convenient to process in software implementations, and today the multiplicity is the more common term in quantum chemistry. Another common nomenclature related to the multiplicity is based on the number of spectral lines (including Zeeman splitting): singlet for $M = 1$, doublet for $M = 2$, triplet for $M = 3$, and so forth.

10.2.3
The Jahn–Teller Effect

It is not always straightforward to predict the multiplicity of a molecule. In the case of a closed-shell molecule with appreciable HOMO–LUMO gap, it is safe to assume that the multiplicity is given by $M = 1$. The situation is different if degenerate, not completely filled, frontier

Figure 10.2 Schematic view of the two energetically degenerate HOMOs of the D_{4h} cyclobutadiene molecule (triplet state, $M = 3$) and the splitting due to a symmetry reduction (D_{2h} symmetry, singlet state, $M = 1$).

orbitals are present. A simple example is cyclobutadiene (C_4H_4). In Chapter 7 (Problem 1) we found cyclobutadiene to be anti-aromatic, that is, with a two-fold degenerate HOMO, which is filled by one electron each (Figure 10.2). As we will show in the Demonstration, we obtain a distorted C_4H_4 with the point group D_{2h} if we optimize the molecule as a closed-shell molecule ($M = 1$). The molecule breaks symmetry, which has the consequence that the HOMO is no longer degenerate. A HOMO–LUMO gap arises and the molecule has a closed-shell electronic structure, though with only a small HOMO–LUMO gap. The situation is different if we request a multiplicity of $M = 3$ (a triplet state). During the geometry optimization the D_{4h} symmetry is maintained. If we compare the energies we find that the distorted D_{2h} isomer with $M = 1$ is more stable. This effect is called Jahn–Teller distortion: the reduction of symmetry removes the degeneracy of the molecular orbitals. The occupied orbital is stabilized, all electrons are paired, and we obtain a more stable structure.

Jahn–Teller distortion

For molecules as simple as cyclobutadiene it is straightforward to predict the existence of a biradicaloid isomer. The situation is different for larger molecules; a nice example is the unstable isomer of the C_{76} fullerene, which was discussed in Chapter 7. If the multiplicity of a molecule is not obvious, a good strategy is to calculate the initial structure with the lowest possible multiplicity. In the case of a large HOMO–LUMO gap, no further considerations are required. If the gap is small, or there are even technical problems to calculate the orbitals, calculations with higher multiplicities should be carried out, and comparison of the total energies of the resulting structures indicates the correct multiplicity.

10.3
Demonstration

Elementary MO theory shows that the oxygen molecule (O_2) can only form a triplet state. The MO degeneracy cannot be lowered by symmetry breaking (because the symmetry of a homoatomic dimer will always be $D_{\infty h}$), and Hund's rule forces the electrons into the triplet configuration. Create the O_2 molecule and try to optimize the geometry. Many approximations to the Schrödinger equation, including those made in the deMon code, fail numerically to solve the Schrödinger equation. If the correct multiplicity is chosen, the molecule converges easily into its equilibrium geometry. Our example code deMon chooses by default the lowest possible multiplicity. Different values can be enforced by the keyword

```
MULTIPLICITY M
```

M being the requested (integer) value for the multiplicity.

Now we investigate the cyclobutadiene molecule. We create the geometry such that (i) the molecule is planar, (ii) the bond angles between the carbons are 90°, and (iii) the carbon ring is a rectangle with slightly different bond lengths (e.g., 1.45 and 1.50 Å, respectively). Hydrogens are attached such that their HCC angle is 135°, so the molecule belongs to the point group D_{2h}. Now we optimize the geometry of the molecule with (a) $M = 1$ and (b) $M = 3$. Inspect the final geometries. For (a), you will observe that the molecule remains in its proposed point group; for (b), the molecule increases symmetry to D_{4h} (within some numerical tolerance).

10.4
Problems

1. Molecular orbital theory for nitrogen monoxide

Draw the MO diagram of nitrogen monoxide (nitric oxide, NO). Perform a quantum chemical calculation and compare the molecular orbitals you obtain with those of the simplistic MO diagram.

2. The Jahn–Teller effect

Calculate the structure of cyclooctatetraene anion ($C_8H_8^-$), starting from D_{8h} geometry. What happens to the geometry? Explain the effect in terms of the Jahn–Teller effect.

3. A stable biradical

Note: The following task is computer-time intensive and might not run on older hardware.

If your computer allows, calculate the relative energy of two isolated triphenylmethyl radicals and the triphenylmethyl dimer in its symmetric and asymmetric form (see Figure 10.1). Visualize the frontier orbital of the radical. Explain why the radicals are relatively stable compared to the dimers of this molecule.

Note: The interaction between the phenyl rings is poorly approximated within density-functional theory.

For deMon, a correction to this method is applied with the following keyword:

DISPERSION

If you encounter numerical difficulties, you may want to activate a numerical trick to improve the self-consistent field (SCF) convergence, which is separating occupied from unoccupied orbitals, the so-called level shift. In deMon, the keyword is

SHIFT 0.2

for a level shift of reasonable 0.2 Hartree.

First, optimize the geometry of the triphenylmethane molecule, which is a closed-shell molecule. Then, form the radical by removing the hydrogen atom from the central carbon and redo the optimization using the correct multiplicity ($M = 2$) to obtain the total energy of the radical. Finally, build each of the two dimers successively with a molecular editor (e.g., Molden) by adding more and more phenyl fragments to a carbon backbone system. The overall computing time is reduced, if you continue to optimize each fragment.

4. Stable molecules formed by unstable fragments

Various radicals are stabilized by forming ions. A typical example is the cyclopentadienyl anion Cp^- ($C_5H_5^-$, D_{5h}). Compare the structure of the neutral species with that of its anion using an electronic structure method and Hückel theory (Chapter 7). One compound in which Cp^- is stabilized by a transition metal is ferrocene, $Fe(C_5H_5)_2$. Build this molecule with a molecular editor using Cp^- building blocks and optimize its geometry.

10.5
Technical Details

If you carry out the problems using deMon, keep the following ideas in mind.
For problem 3 always use the additional keyword

ERIS DIRECT

as the molecules get too large for the conventional SCF technique[1]. If you have problems in SCF convergence, use a level shift (keyword SHIFT LS, LS being the level shift in Hartree, see above). It is always a good idea to check if a calculation that has been converged using a level

---

1) In conventional SCF, all integrals with the basis functions are calculated and stored in memory or on disk. For large systems, the memory is either insufficient, or it is ineffective to store these data. Instead, direct SCF recalculates integrals whenever they are needed.

shift has really reached the ground state. For this purpose, restart the converged calculation, but remove the level shift keyword and check the energy and the molecular orbitals afterwards. This is achieved by using the keyword

    GUESS RESTART

Whenever radicals are processed, it is a good idea to monitor the orbital energies. In deMon, use the keyword

    PRINT MOE

10.6
Review and Summary

You have learned that molecules containing unpaired electrons require special attention, not only in experimental but also in computational chemistry. The treatment of radicals and other species with non-zero net spin is more tedious and sometimes involves technical difficulties, but it is straightforward in most computational chemistry codes. Most approaches treat spin-up and spin-down electrons independently and thus cost twice the computer time and memory compared to a calculation of a molecule of the same size, but with zero net spin (multiplicity 1).

Radicals may occur in different situations, for example (i) if the number of electrons in the molecule is odd, (ii) if the molecule contains elements with incomplete f shells, or (iii) if symmetry imposes degeneracy of the frontier orbitals. Some radicals might be quite stable, in particular if the unpaired electron does not participate in the bonding (as for some elements with an incomplete f shell) or if it can be delocalized, for example, in aromatic rings.

For large molecules, it is not obvious if a radicaloid character is possible. In these cases, the HOMO–LUMO gap should be inspected; if it is small the calculation should be repeated using a different multiplicity. Monitoring the MO energies is always beneficial; this is true in particular if a calculation does not converge because of a wrong electron occupation. We must not forget that changing the multiplicity changes the structure of the molecule. Therefore, geometry optimizations have to be carried out within the requested spin state.

Reference

1 For the development of this important law of quantum chemistry, see Wolfgang Pauli's Nobel lecture (http://nobelprize.org/nobel_prizes/physics/laureates/1945/pauli-lecture.html).

11
Vibrational Spectroscopy

11.1
Aim

In this computer experiment, you will calculate infrared (IR) and Raman spectra. You will start with the calculation of the vibrational frequencies of the CO_2 molecule by hand, interpret your results, and compare them to experimental data. Later, you will calculate IR and Raman spectra using modern tools of computational chemistry, and you will visualize the vibrational modes and frequencies with graphical software.

11.2
Theoretical Background

Infrared and Raman spectroscopy are important methods for the characterization of molecules [1, 2]. The possibility of calculating an IR spectrum is very helpful for the interpretation of experimental work, because the calculated spectrum allows a direct assignment of the vibrational frequency[1] and mode. Especially in combination with nuclear magnetic resonance (NMR) techniques, vibrational spectroscopy allows the molecular structure of an unknown substance to be determined by comparing the calculated and experimentally measured spectra without performing X-ray diffraction analyses, which require the presence of crystalline material in appreciable quantities. Since the IR and Raman absorption occurs due to molecular vibrations, we speak also of vibrational spectroscopy.

vibrational frequency

IR spectroscopy
Raman spectroscopy

vibrational spectroscopy

---

[1] Note that in IR spectroscopy, frequencies are usually given in terms of wave numbers $\bar{v}$ (in cm^{-1}), which are reciprocal wavelengths λ. Frequencies f (often also named v) are connected to both by the expressions $f = c/\lambda = c\,\bar{v}$. All three quantities express the energy of a vibration in different ways, and they can be converted into each other following: $E = hf = hc/\lambda = hc\,\bar{v}$.

Computational Chemistry Workbook: Learning Through Examples
Thomas Heine, Jan-Ole Joswig, and Achim Gelessus
Copyright © 2009 WILEY-VCH Verlag GmbH & Co. KGaA, Weinheim
ISBN: 978-3-527-32442-2

11.2.1
Analysis of Classical Vibrations Within the Harmonic Approximation

vibrational mode
normal mode

The vibrations of a molecule are defined through their vibrational modes or normal modes, which have been associated with vibrational energy levels (see Chapter 4). In a vibrational spectrum, each absorption is associated with a vibrational mode. Figure 11.1 shows the normal modes of the linear carbon dioxide molecule: the symmetric, the anti-symmetric, and the two bending modes. The latter two have the same energy, but differ in the direction of the vibration. Vibrational modes that have the same energy are called degenerate.

In the following we will use the harmonic approximation (see Chapter 3), which is the most common approximation to calculate vibrational frequencies and modes. Linear N-atomic molecules have $3N - 5$ vibrational modes. The five remaining degrees of freedom are three for the linear translation of the molecular center of mass in space and two more degrees of freedom for the rotation along the molecular axis and perpendicular to it. Nonlinear molecules have three rotational degrees of freedom and thus $3N - 6$ vibrational modes. Within the harmonic approximation, the normal modes have the following properties.

degrees of freedom

1. Each vibrational mode acts like an independent harmonic oscillator.
2. Each vibration contains motions of many atoms, usually of all.
3. The molecular center of mass stays at rest.
4. All atoms pass through their equilibrium position at the same time.
5. All vibrational modes are independent of each other.

In the antisymmetric stretching mode and the two bending modes of the carbon dioxide molecule, all the atoms move. The collective movement of atoms is a generic property of vibrational modes. In contrast,

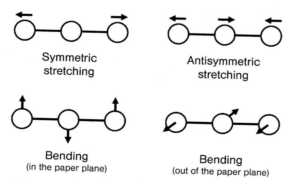

Symmetric
stretching

Antisymmetric
stretching

Bending
(in the paper plane)

Bending
(out of the paper plane)

Figure 11.1 Normal modes of a linear triatomic molecule, for example, carbon dioxide. The two bending modes are energetically degenerate and differ only in the direction of deformation (in and out of the paper).

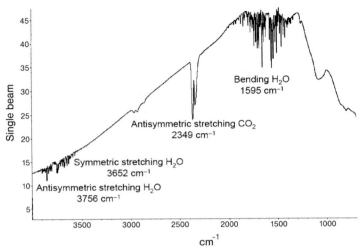

Figure 11.2 The infrared spectrum of the atmosphere. The presented spectrum shows the background absorption of an FT-IR spectrometer. (Courtesy of Prof. Dr. Ulrich Kortz and Dr. Bassem Bassil, Jacobs University Bremen).

the central atom stays at rest in the symmetric stretching mode. This is necessary to keep the center of mass at rest and is characteristic for molecules with a center of inversion that is located at the position of an atom. Since the vibrational modes are independent of each other, they do not interchange energy. This means that, if a symmetric bending vibration is excited, the energy remains in this vibrational mode. No other mode can be excited from it.

The background spectrum of the atmosphere (Figure 11.2) shows – among other things – the symmetric and antisymmetric stretching modes of water, as well as the antisymmetric stretching and bending modes of carbon dioxide. The absorptions of these molecules are responsible for the greenhouse effect. In contrast, the symmetric stretching mode of CO_2 is not present in the infrared spectrum, because it is visible only in a Raman spectrum. In the following, we will use CO_2 as a simple example to analyze molecular vibrations.

infrared spectrum
Raman spectrum

11.2.2
The Harmonic Oscillator Revisited

We will now briefly recapitulate the concept of the harmonic oscillator, which has already been introduced in Chapter 3. First, we will deal with a one-dimensional system, and later expand this to a multi-dimensional problem. The force F that is exerted on a particle of mass m, which is connected to a spring with force constant k, is defined by Hooke's law:

harmonic oscillator

$F = -k\tilde{X}$. Note that we will use $\tilde{X}$ for the displacement of the atom from its equilibrium position X_e in the following to remain compatible with the nomenclature in Chapter 3. Since the force acts in the opposite direction to the displacement, the force is negative. This means that, if the particle is displaced, the force tries to bring it back to its original equilibrium position. The same is true for a compressed spring. Then $\tilde{X}$ is negative ($\tilde{X} < 0$) and the force is positive, acting in the direction towards the equilibrium position. The potential energy can be calculated by integrating Hooke's law,

$$F = -\frac{dV}{d\tilde{X}} = -k\tilde{X} \tag{11.1}$$

resulting in

$$V = \frac{1}{2}k\tilde{X}^2 = \frac{1}{2}\frac{d^2V}{d\tilde{X}^2}\bigg|_{\tilde{X}=0} \tilde{X}^2 \tag{11.2}$$

Here, we have already replaced the force constant by the second derivative of the potential energy with respect to the displacement.

Combining Hooke's law (Equation 11.1) and Newton's law ($F = ma$) we obtain the result

$$m\underbrace{\frac{d^2\tilde{X}}{dt^2}}_{a} = -k\tilde{X} \tag{11.3}$$

which states that the force (mass times acceleration; left-hand side) equals the (negative) spring constant k times displacement $\tilde{X}$ (right-hand side). In other words, the acceleration back to the equilibrium position depends on the mass, the spring constant, and the extent of the displacement. We would expect that from our everyday experience.

The solutions are known from physics, being the time-dependent solutions of the harmonic oscillator:

$$\tilde{X}(t) = A\sin(\omega t) = A\sin(2\pi f t) \tag{11.4}$$

Here, the vibrational frequency is denoted as f, and the amplitude of the vibration as A. The second derivative[2] with respect to the time results in

---

2) The first derivative of Equation 11.4 is

$$\frac{d\tilde{X}}{dt} = A\cos(2\pi f t) \times 2\pi f$$

The second derivative can then easily be calculated as:

$$\frac{d^2\tilde{X}}{dt^2} = -(2\pi f)^2 \underbrace{A\sin(2\pi f t)}_{\tilde{X}}$$
$$= -4\pi^2 f^2 \tilde{X}.$$

$$\frac{d^2\tilde{X}}{dt^2} = -4\pi^2 f^2 \tilde{X} \tag{11.5}$$

and combined with Equation 11.3 we obtain the result

$$m \times \underbrace{(-4\pi^2 f^2 \tilde{X})}_{d^2\tilde{X}/dt^2} = -k\tilde{X} \tag{11.6}$$

The displacement $\tilde{X}$ cancels and the frequency can be calculated. The frequency is thus dependent on the mass and the force constant only.

In multi-dimensional problems, we have a set of $3N$ coordinates in an N-atomic molecule; Equation 11.6 therefore converts into an eigenvalue problem. The scalar displacement $\tilde{X}$ consequently converts into a vector that contains all $3N$ atomic displacements. Finally, the force constant k of a simple spring converts into a **force-constant matrix** of **force-constant matrix** dimension $3N \times 3N$. The resulting equation is the basis for calculating the vibrational modes of molecules.

11.2.3
The Vibrational Modes

Each atom in a molecule can be described by three Cartesian coordinates, for example, atom 1 is positioned at $(X_{1,e}, Y_{1,e}, Z_{1,e})$, atom 2 at $(X_{2,e}, Y_{2,e}, Z_{2,e})$ and so forth. The index states the number of the atom and shows that these are the equilibrium positions (e). As we have seen already in Chapter 3, it is sometimes useful to replace the absolute coordinates with relative coordinates with respect to the equilibrium position, that is, the displacements of the atoms in all three spatial directions:

atom 1 : $\quad \tilde{X}_1 = X_1 - X_{1,e}, \quad \tilde{Y}_1 = Y_1 - Y_{1,e} \quad \tilde{Z}_1 = Z_1 - Z_{1,e}$

atom 2 : $\quad \tilde{X}_2 = X_2 - X_{2,e}, \quad \tilde{Y}_2 = Y_2 - Y_{2,e} \quad \tilde{Z}_2 = Z_2 - Z_{2,e}$

$\qquad \vdots \qquad\qquad \vdots \qquad\qquad \vdots \qquad\qquad \vdots$

atom i : $\quad \tilde{X}_I = X_I - X_{I,e}, \quad \tilde{Y}_I = Y_I - Y_{I,e} \quad \tilde{Z}_I = Z_I - Z_{I,e}$

$$\tag{11.7}$$

For the general case of atom i, the atom is in its equilibrium position $(X_{I,e}, Y_{I,e}, Z_{I,e})$ if the displacement $\tilde{X}_I, \tilde{Y}_I$ and $\tilde{Z}_I$ are zero in each direction.

In the Born–Oppenheimer approximation, the potential energy of an **Born–Oppenheimer** N-atomic molecule is a function of the atomic positions, that is, the **approximation** nuclear coordinates: $V(X_1, Y_1, Z_1, X_2, Y_2, Z_2, \ldots, X_N, Y_N, Z_N)$. The second derivative of the potential energy with respect to the coordinates can now be used to calculate the force constant in Equation 11.2. One just has to take care of the fact that the force constant is now a matrix, that is, the potential energy depends on all $3N$ coordinates of the N-atomic molecule, so that we have to derive the potential energy with respect to all $3N$ coordinates. As a result, we will get $3N$ first derivatives, and each of these has to be derived a second time with respect to all $3N$ coordinates. We end up with a $3N \times 3N$ matrix containing $(3N)^2$ second derivatives. The X component of the force vector acting on atom 1, for example, changes as

$$\frac{\partial^2 V}{\partial \tilde{X}_1^2} = k_{XX}^{11} \tag{11.8}$$

if the atom is displaced in the X direction. If atom 1 is displaced in the Y direction, in contrast, the X component of the force acting on the atom changes as

$$\frac{\partial^2 V}{\partial \tilde{Y}_1 \partial \tilde{X}_1} = k_{XY}^{11} \tag{11.9}$$

Note the order of differentiation. As an example, Figure 11.3 shows the different types of force constants.

The force-constant matrix contains not the force constants of specific bonds (as for the diatomic molecule), but the influence of the displacement of an atom on the force components of another (or the same) atom decomposed into the three spatial contributions. Thereby, it is not important if the two atoms are connected by a chemical bond or not. The

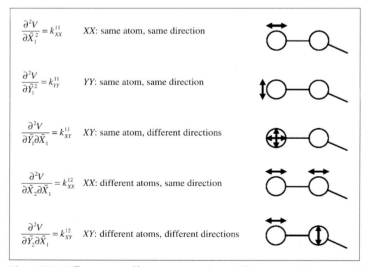

Figure 11.3 Different types of force constants, that is, different second derivatives of the potential energy.

complete list of force constants is usually written as a $3N \times 3N$ matrix containing $(3N)^2$ matrix elements. This force-constant matrix is called the Hesse matrix (or Hessian).   **Hesse matrix**

For each force-constant matrix element, we will now write down Equation 11.6:

$$-4\pi^2 f^2 m_1 \tilde{X}_1 = -k_{XX}^{11}\tilde{X}_1 - k_{XY}^{11}\tilde{Y}_1 - k_{XZ}^{11}\tilde{Z}_1 - k_{XX}^{12}\tilde{X}_2 - k_{XY}^{12}\tilde{Y}_2 - \cdots - k_{XZ}^{1N}\tilde{Z}_N$$

$$-4\pi^2 f^2 m_1 \tilde{Y}_1 = -k_{YX}^{11}\tilde{X}_1 - k_{YY}^{11}\tilde{Y}_1 - k_{YZ}^{11}\tilde{Z}_1 - k_{YX}^{12}\tilde{X}_2 - k_{YY}^{12}\tilde{Y}_2 - \cdots - k_{YZ}^{1N}\tilde{Z}_N$$

$$\vdots$$

$$-4\pi^2 f^2 m_2 \tilde{X}_2 = -k_{XX}^{21}\tilde{X}_1 - k_{XY}^{21}\tilde{Y}_1 - k_{XZ}^{21}\tilde{Z}_1 - k_{XX}^{22}\tilde{X}_2 - k_{XY}^{22}\tilde{Y}_2 - \cdots - k_{XZ}^{2N}\tilde{Z}_N$$

$$\vdots$$

$$-4\pi^2 f^2 m_N \tilde{Z}_N = -k_{ZX}^{N1}\tilde{X}_1 - k_{ZY}^{N1}\tilde{Y}_1 - k_{ZZ}^{N1}\tilde{Z}_1 - k_{ZX}^{N2}\tilde{X}_2 - k_{ZY}^{N2}\tilde{Y}_2 - \cdots - k_{ZZ}^{NN}\tilde{Z}_N$$

$$(11.10)$$

The right-hand side of each equation shows that the total force acting on atom I is the sum of the force components of the forces exerted on atom I by all other atoms. Since each displacement has three spatial components, there are $3N$ equations to solve. Each of the $3N$ equations has $3N$ terms on its right-hand side. The solutions to this eigenvalue problem result in $3N$ eigenvalues and $3N$ corresponding eigenvectors. The eigenvalues of this eigenvalue problem are related to the vibrational frequencies, and the eigenvectors describe the corresponding vibrational modes. A simplified eigenvalue problem will be solved in the Demonstration.

11.2.4
Intensities

Besides the frequencies, the IR and Raman intensities can also be obtained from quantum chemical calculations. However, we will discuss these only briefly and phenomenologically. The IR intensity is proportional to the squared change of the dipole moment of the molecule during the vibration. A large dipole moment change causes a large intensity, and vice versa. As a consequence, the IR spectrum contains only those vibrations during which the molecular dipole moment changes with time. This is the reason why in Figure 11.2 the symmetric stretching mode of CO_2 is not present: in this mode the molecular dipole moment does not change.

The Raman intensities are caused by a change in the polarizibilities of the molecule during the vibration, and only those vibrations during which the polarizabilities change are observed in a Raman spectrum. If the intensity of a vibration is non-zero, the vibration is called infrared-active or Raman-active. Inactive vibrations have zero intensity and thus are not observed.

11.3
Demonstration

11.3.1
The Vibrational Modes of a Linear Molecule

In the following we will analyze the vibrational modes of a linear triatomic molecule as an example, and as before we choose carbon dioxide. The molecule is linear, and we restrict our investigation for the moment to only those vibrations which are aligned with the X axis. There are more vibrations, for example the bending of the molecule, which can be treated if the X and Y components of the coordinates are included. For our demonstration example, however, we will restrict the investigation to a one-dimensional problem. Therefore, we can define a coordinate system along the molecular axis (Figure 11.4), and Equations 11.10 reduce to three equations only:

$$-4\pi^2 f^2 m_1 \tilde{X}_1 = -k_{XX}^{11}\tilde{X}_1 - k_{XX}^{12}\tilde{X}_2 - k_{XX}^{13}\tilde{X}_3$$

$$-4\pi^2 f^2 m_2 \tilde{X}_2 = -k_{XX}^{21}\tilde{X}_1 - k_{XX}^{22}\tilde{X}_2 - k_{XX}^{23}\tilde{X}_3 \qquad (11.11)$$

$$-4\pi^2 f^2 m_3 \tilde{X}_3 = -k_{XX}^{31}\tilde{X}_1 - k_{XX}^{32}\tilde{X}_2 - k_{XX}^{33}\tilde{X}_3$$

There are different possibilities of solving Equation 11.11. The usual way, which is also implemented in most computational codes, will be discussed in the following. As a first step, we introduce mass-weighted coordinates for the three atoms 1, 2, and 3,

$$\bar{X}_1 = \sqrt{m_1}\tilde{X}_1, \quad \bar{X}_2 = \sqrt{m_2}\tilde{X}_2, \quad \bar{X}_3 = \sqrt{m_3}\tilde{X}_3 \qquad (11.12)$$

and mass-weighted force-constant matrix elements

$$\bar{k}_{XX}^{ij} = \frac{k_{XX}^{IJ}}{\sqrt{m_I m_J}} \qquad (11.13)$$

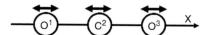

Figure 11.4 Schematic view of the CO_2 molecule including the numbering of the atoms. Only the displacements along the molecular (X) axis are considered.

With the introduction of the mass-weighted coordinates and force constants, Equations 11.11 can be simplified to:

$$-4\pi^2 f^2 \bar{X}_1 = -\bar{k}_{XX}^{11}\,\bar{X}_1 - \bar{k}_{XX}^{12}\,\bar{X}_2 - \bar{k}_{XX}^{13}\,\bar{X}_3$$
$$-4\pi^2 f^2 \bar{X}_2 = -\bar{k}_{XX}^{21}\,\bar{X}_1 - \bar{k}_{XX}^{22}\,\bar{X}_2 - \bar{k}_{XX}^{23}\,\bar{X}_3 \qquad (11.14)$$
$$-4\pi^2 f^2 \bar{X}_3 = -\bar{k}_{XX}^{31}\,\bar{X}_1 - \bar{k}_{XX}^{32}\,\bar{X}_2 - \bar{k}_{XX}^{33}\,\bar{X}_3$$

This set of three equations can also be written as a matrix equation, the matrix being the Hesse matrix of our model system, the linear carbon dioxide molecule:

$$-\begin{pmatrix} \bar{k}_{XX}^{11} & \bar{k}_{XX}^{12} & \bar{k}_{XX}^{13} \\ \bar{k}_{XX}^{21} & \bar{k}_{XX}^{22} & \bar{k}_{XX}^{23} \\ \bar{k}_{XX}^{31} & \bar{k}_{XX}^{32} & \bar{k}_{XX}^{33} \end{pmatrix} \begin{pmatrix} \bar{X}_1 \\ \bar{X}_2 \\ \bar{X}_3 \end{pmatrix} = -4\pi^2 f^2 \begin{pmatrix} \bar{X}_1 \\ \bar{X}_2 \\ \bar{X}_3 \end{pmatrix} \qquad (11.15)$$

Symmetry requires that the Hesse matrix itself is symmetric with respect to the diagonal, for example, $\bar{k}_{XX}^{31} = \bar{k}_{XX}^{13}$. This fact reduces the number of unknown matrix elements from nine to six (three diagonal elements and three off-diagonal elements).

Finally, we perform a last and simple transformation by dividing both sides by $-4\pi^2 f^2$:

$$\frac{1}{4\pi^2 f^2} \begin{pmatrix} \bar{k}_{XX}^{11} & \bar{k}_{XX}^{12} & \bar{k}_{XX}^{13} \\ \bar{k}_{XX}^{21} & \bar{k}_{XX}^{22} & \bar{k}_{XX}^{23} \\ \bar{k}_{XX}^{31} & \bar{k}_{XX}^{32} & \bar{k}_{XX}^{33} \end{pmatrix} \begin{pmatrix} \bar{X}_1 \\ \bar{X}_2 \\ \bar{X}_3 \end{pmatrix} = \begin{pmatrix} \bar{X}_1 \\ \bar{X}_2 \\ \bar{X}_3 \end{pmatrix} \qquad (11.16)$$

This is a special eigenvalue problem, which can solved by hand or with the eigenvalue solver on the CD, which is distributed with this book. The obtained eigenvalues are the squared vibrational frequencies. If the investigated system is not in its total energy minimum, but in a transition state, one (or more) negative eigenvalues will occur, because in the transition state the potential energy surface is flat (first derivative equals zero), but its curvature (second derivative) in different directions is positive and negative at the same time. Negative eigenvalues from Equation 11.16 will lead to imaginary frequencies.

We will now use approximate force constants to calculate the vibrational frequencies in the simplified coordinate system of Figure 11.4. The force constant for the C$-$O bond is approximately 1600 N m^{-1}. The corresponding force-constant matrix elements for the displacement of the two oxygen atoms are $k_{XX}^{11} = k_{XX}^{33} = 1600$ N m^{-1}. Since the carbon atom has two bonds, the resulting matrix element is $k_{XX}^{22} = 2k_{XX}^{11} = 3200$ N m^{-1}. The interactions between the two oxygen atoms are neglected, that is, $k_{XX}^{13} = k_{XX}^{31} = 0$. For $k_{XX}^{12} = k_{XX}^{21} = k_{XX}^{23} = k_{XX}^{32}$ we use the force-constant matrix element k_{XX}^{11}, but with opposite sign.

For k_{XX}^{11}, atom 1 is displaced towards atom 2 (with respect to the coordinate system introduced in Figure 11.4); for k_{XX}^{12}, atoms 1 and 2 are moving simultaneously. The resulting force constant matrix looks as follows:

$$\boldsymbol{k} = \begin{pmatrix} k_{XX}^{11} & k_{XX}^{12} & k_{XX}^{13} \\ k_{XX}^{21} & k_{XX}^{22} & k_{XX}^{23} \\ k_{XX}^{31} & k_{XX}^{32} & k_{XX}^{33} \end{pmatrix} = \begin{pmatrix} 1600 & -1600 & 0 \\ -1600 & 3200 & -1600 \\ 0 & -1600 & 1600 \end{pmatrix} \text{Nm}^{-1}$$

$$(11.17)$$

We can interpret the approximated force-constant matrix given in Equation 11.17 as follows. Moving atom O^3 has no influence on the force constant of atom $O^1(k_{XX}^{13} = 0)$, whereas moving atom C^2 has an influence on both oxygen atoms ($k_{XX}^{12} = k_{XX}^{23} \neq 0$). The influence of the displacement of an atom on its own force constant is different for the two oxygen atoms and the central carbon atom, because the latter is bonded to two atoms.

Mass weights are now introduced according to Equation 11.13. The molar masses of carbon and oxygen are $M_C = 12$ g mol^{-1} and $M_O = 16$ g mol^{-1} (we may use approximate values here as well) and the mass is $m = M/N_A$. So we obtain

$$\bar{\boldsymbol{k}} = \begin{pmatrix} \dfrac{1600}{\sqrt{16}\sqrt{16}} & \dfrac{-1600}{\sqrt{16}\sqrt{12}} & 0 \\ \dfrac{-1600}{\sqrt{16}\sqrt{12}} & \dfrac{3200}{\sqrt{12}\sqrt{12}} & \dfrac{-1600}{\sqrt{12}\sqrt{16}} \\ 0 & \dfrac{-1600}{\sqrt{12}\sqrt{16}} & \dfrac{1600}{\sqrt{16}\sqrt{16}} \end{pmatrix} \text{N mol m}^{-1}\text{g}^{-1} \times N_A$$

$$(11.18)$$

The total eigenvalue problem of Equation 11.16, which is a special eigenvalue problem that can be solved using the eigenvalue solver on the CD, becomes

$$\frac{\text{N mol m}^{-1}\text{g}^{-1} \times N_A}{4\pi^2 f^2} \begin{pmatrix} 100 & -115.47 & 0 \\ -115.47 & 266.67 & -115.47 \\ 0 & -115.47 & 100 \end{pmatrix} \begin{pmatrix} \bar{X}_1 \\ \bar{X}_2 \\ \bar{X}_3 \end{pmatrix} = \begin{pmatrix} \bar{X}_1 \\ \bar{X}_2 \\ \bar{X}_3 \end{pmatrix}$$

$$(11.19)$$

The results are presented in Table 11.1 and can be easily interpreted. The first eigenvalue is nearly zero, and the corresponding eigenvector shows a displacement of all three atoms in the same direction. This is a translational degree of freedom and therefore not relevant for our problem, though a solution of the eigenvalue problem. It has the characteristic eigenvalue of 0, indicating that the energy of the molecule is not changing upon moving the molecule along this mode: the

Table 11.1 Eigenvalues and eigenvectors of the eigenvalue problem in Equation 11.16. The input for the eigenvalue solver is just the matrix $\tilde{k}$. The units and the factor $N_A/4\pi^2 f^2$ will be included, when calculating the frequencies.

No.	Eigenvalue $(N\,mol\,m^{-1}\,g^{-1})$	Eigenvector		
		1	2	3
1	0.0010	−0.6030	−0.5222	−0.6030
2	100.0000	−0.7071	0.0000	0.7071
3	366.6690	0.3693	−0.8528	0.3693

molecular geometry remains fixed. Note that the different values of the eigenvector components result from the mass-weighted coordinates that have been used in Equation 11.19. In the second vibrational mode, the carbon atom stays at rest and the two oxygen atoms are displaced by the same amount in opposite directions, which is a symmetric stretching mode. Finally, the last mode shows the two oxygen atoms moving in the same direction, whereas the carbon atom is moving towards one oxygen atom. This is the antisymmetric stretching mode. We notice that bending modes are absent from our solution. This is not surprising, since we have restricted ourselves to the linear coordinate system (and only linear vibrational modes) in Figure 11.4.

As a final step we have to calculate the frequencies from the eigenvalues. We will do that only for eigenvalues nos. 2 and 3, which are, in contrast to eigenvalue no. 1, vibrational modes. We know from Chapter 3 (Demonstration) that the wavenumber is $\bar{v} = \omega_0/2\pi c$ and that $\omega_0 = \sqrt{k/m}$. These two relations can be combined to result in

$$\frac{4\pi^2 c^2 \bar{v}^2}{N_A} = \frac{k}{M} \quad \text{and} \quad \bar{v} = \sqrt{\frac{N_A}{4\pi^2 c^2}\frac{k}{M}} \tag{11.20}$$

whereby we have squared the expression in order to remove the square root, and we used $m = M/N_A$. Note that the eigenvalues in Table 11.1 are given in $N\,mol\,m^{-1}\,g^{-1}$, whereas we use $N\,mol\,m^{-1}\,kg^{-1}$ to calculate the frequencies:

$$\bar{v}_2 = \sqrt{\frac{6.022 \times 10^{23}\,mol^{-1}}{4\pi^2 \times (2.998 \times 10^8\,ms^{-1})^2} 100.0 \times 10^3 \frac{N\,mol}{m\,kg}}$$
$$= 130274\,m^{-1} = 1302.74\,cm^{-1}$$

$$\bar{v}_3 = \sqrt{\frac{6.022 \times 10^{23}\,mol^{-1}}{4\pi^2 \times (2.998 \times 10^8\,ms^{-1})^2} 366.669 \times 10^3 \frac{N\,mol}{m\,kg}}$$
$$= 249457\,m^{-1} = 2494.57\,cm^{-1}$$

Table 11.2 Experimental data for the vibrations of CO_2 and HCN [3].

Molecule	Mode	Wavenumber (cm^{-1})
CO_2	Symmetric stretching	1333
	Antisymmetric stretching	2349
	Bending	667
HCN	HC stretching	3311
	CN stretching	2097
	Bending	712

The resulting vibrational frequencies (in terms of wavenumbers) are thus $\bar{v}_2 = 1302.74 \text{ cm}^{-1}$ and $\bar{v}_3 = 2494.57 \text{ cm}^{-1}$. The corresponding experimental values are 1333 cm^{-1} and 2349 cm^{-1} (see Table 11.2). Although we have made severe approximations, we were able to reproduce the two stretching frequencies with errors of just 2.8% and 6.2%, respectively.

We can also see that an eigenvalue of 0 results in a wavenumber (and energy) of 0. The translational mode (eigenvalue 1) is therefore no vibration.

11.4
Problems

1. Calculation of the frequencies of the HCN molecule

Calculate the vibrational frequencies and the vibrational modes of the hydrogen cyanide (HCN) molecule by hand as shown in the Demonstration. Consider only the stretching modes, and use the simplification of the coordinate system as in Figure 11.4. Use the following force constants: 622 N m^{-1} for the H−C bond, and 1878 N m^{-1} for the C−N bond. Use the following values for the corresponding force-constant matrix elements for the displacement of the H and N atom, and their sum for the matrix element of the C atom (since it has two bonds): $k_{XX}^{11} = 622 \text{ N m}^{-1}$, $k_{XX}^{22} = 1878 \text{ N m}^{-1}$ and $k_{XX}^{22} = k_{XX}^{11} + k_{XX}^{33}$. The interactions between the hydrogen and the nitrogen atoms are neglected, that is, $k_{XX}^{13} = k_{XX}^{31} = 0$. For $k_{XX}^{12} = k_{XX}^{21}$, we use the force-constant matrix element k_{XX}^{11}, and for $k_{XX}^{23} = k_{XX}^{32}$, we use the force-constant matrix element k_{XX}^{11}, but both with opposite sign. For k_{XX}^{11}, atom 1 is displaced towards atom 2 (with respect to the coordinate system introduced in Figure 11.4), and for k_{XX}^{12}, atoms 1 and 2 are moving simultaneously. (Note that these force constants are not mass-weighted.) Sketch the vibrational modes graphically.

2. Electronic structure calculation of the frequencies of the CO_2 molecule

Calculate the IR and Raman spectra of the carbon dioxide (CO_2) and the hydrogen cyanide (HCN) molecules using an electronic structure method, for example, density-functional theory as implemented in the `deMon` code, which is available on the CD distributed with this book. Compare your results with those obtained in the Demonstration and Problem 1, as well as with available experimental data. Visualize the vibrational modes in a molecular editor (e.g., `Molden`). Which modes have been neglected in problem 1? Which CO_2 and HCN modes are infrared-active, and which are Raman-active? Why?

3. Infrared and Raman spectra of different molecules

Calculate the IR and Raman spectra of water (H_2O), ammonia (NH_3), and pyrrole using an electronic structure method, for example, density-functional theory as implemented in the `deMon` code. Visualize the vibrational modes. Do you expect different vibrational modes (stretching or bending modes) to occur in different frequency ranges? Explain your answer.

4. Infrared and Raman spectra of difluoroethene

Calculate the IR and Raman spectra of *cis*- and *trans*-difluoroethene using an electronic structure method, for example, density-functional theory as implemented in the `deMon` code. How may your calculated spectra be used to distinguish the two molecules in an experiment?

11.5
Technical Details

The eigenvalue problem in problem 1 can be solved either by hand or by using the eigenvalue solver provided on the CD (or at www.compchem.jacobs-university.de).

If you use `deMon` as a quantum chemical tool for the problems, then first perform a geometry optimization of your system. The optimized structure can be found, for example, in the output file at `FINAL INPUT ORIENTATION`. The frequency calculations can be performed with the optimized molecular coordinates. The `deMon` keyword `FREQUENCY` is needed to calculate the IR spectrum of the system of interest. With the keyword `FREQUENCY RAMAN`, both the IR and the Raman spectra are calculated.

Experimental reference data can be found in numerous databases, for example, at the National Institute of Standards and Technology (http://webbook.nist.gov/chemistry/) or the National Institute of

Advanced Industrial Science and Technology (AIST) (http://riodb01.ibase.aist.go.jp/sdbs).

In `Molden`, the **Norm. Mode** button displays the vibrational modes. Choose a frequency by clicking on it. If `molden` reports negative frequency values, they need to be interpreted as imaginary frequencies.

11.6
Review and Summary

In this chapter, you have been introduced to calculating the infrared-active and Raman-active vibrational modes of molecules both by hand and using modern electronic structure methods. This is a common task in computational chemistry, since IR and Raman spectroscopy are important methods for the characterization of molecules. The vibrational frequencies are characteristic for different molecular structures. Therefore, IR and Raman spectroscopy can be used to determine the structure of a molecule with the help of computational chemistry. You have also learned how to visualize vibrational modes with a molecular editor.

We have used the harmonic approximation, which has been briefly revisited, to describe the displacements of the atoms during the vibrations of a molecule. Thereby, we note that each vibrational mode acts like an independent harmonic oscillator, each vibration contains motions of many (usually of all) atoms, the molecular center of mass stays at rest, all atoms pass through their equilibrium position at the same time, and all vibrational modes are independent of each other. Moreover, we have examined the symmetric and antisymmetric stretching modes as well as the two bending modes of the carbon dioxide molecule.

Finally, we have introduced the relative coordinates $\tilde{X}$, $\tilde{Y}$, and $\tilde{Z}$ to describe the harmonic vibration. With these, the Hessian matrix has been built, which is the force-constant matrix stating the influence of the displacement of a particular atom on the force components of another (or the same) atom, and we obtain an eigenvalue problem. In the Demonstration we have solved a simplified eigenvalue problem for the linear CO_2 molecule and obtained the vibrational frequencies with reasonable accuracy.

References

1 Wilson, E.B., Decius, J.C. and Cross, P.C. (1955) *Molecular Vibrations: The Theory of Infrared and Raman Vibrational Spectra*, McGraw-Hill, New York.

2 Atkins, P.W. and de Paula, J. (2006) *Physical Chemistry*, 4th edn, Oxford University Press.

3 Lide, D.R. (ed.) (2008) *CRC Handbook of Chemistry and Physics*, 89th edn CRC Press, Boca Raton, FL.

12
Vibrational Spectroscopy and Character Tables – Advanced Topics

12.1
Aim

This chapter is devoted to students who have a strong interest in vibrational spectroscopy. We will not only calculate the vibrational spectrum of a molecule, but also analyze the individual vibrational modes. A general procedure to analyze the vibrations in polyatomic molecules, expressed by the normal modes, is presented. For symmetric molecules, symmetry is also an inherent feature of the normal modes. For this purpose, a phenomenological introduction to irreducible representations, character tables and direct products is given. Selection rules for infrared and Raman spectroscopy are derived and discussed briefly.

vibrational spectroscopy

12.2
Theoretical Background

12.2.1
The Hessian Matrix

So far, we have used the Cartesian coordinates of a molecule as a set of vectors pointing at the atomic positions. In this chapter, it is convenient to use a single-vector representation of the Cartesian coordinates, a long vector U that contains the components of the Cartesian position vectors

Computational Chemistry Workbook: Learning Through Examples
Thomas Heine, Jan-Ole Joswig, and Achim Gelessus
Copyright © 2009 WILEY-VCH Verlag GmbH & Co. KGaA, Weinheim
ISBN: 978-3-527-32442-2

R_I of each atom in the molecule:

$$\{R_I\} = R_1, R_2, \ldots, R_N$$

$$= \begin{pmatrix} X_1 \\ Y_1 \\ Z_1 \end{pmatrix}, \begin{pmatrix} X_2 \\ Y_2 \\ Z_2 \end{pmatrix}, \ldots, \begin{pmatrix} X_N \\ Y_N \\ Z_N \end{pmatrix} \rightarrow \begin{pmatrix} X_1 \\ Y_1 \\ Z_1 \\ X_2 \\ Y_2 \\ Z_2 \\ \vdots \\ X_N \\ Y_N \\ Z_N \end{pmatrix} = \begin{pmatrix} U_1 \\ U_2 \\ U_3 \\ U_4 \\ U_5 \\ U_6 \\ \vdots \\ U_{3N-2} \\ U_{3N-1} \\ U_{3N} \end{pmatrix} = U$$

For an N-atomic molecule, the potential energy surface around the equilibrium position can be expanded in a Taylor series in terms of the $3N$ Cartesian coordinates U_i. Here, the first derivatives of the potential with respect to the Cartesian coordinates vanish (since we are at the energy minimum), and if we neglect third and higher order derivatives we obtain the harmonic approximation for polyatomic molecules:

$$V(U) = \frac{1}{2} \sum_i^{3N} \sum_j^{3N} \frac{\partial^2 V(U)}{\partial U_i \, \partial U_j}_{\{U=U_e\}} (U_i - U_{i,e})(U_j - U_{j,e}) \quad (12.1)$$

The partial derivatives in Equation 12.1 are with respect to all $3N$ Cartesian coordinates at the equilibrium geometry. As introduced in Chapter 11, the second derivatives of the potential energy surface with respect to the Cartesian coordinates form a symmetric $3N \times 3N$ matrix **Hessian matrix** (the Hessian matrix) with $(3N)^2$ elements. The further treatment follows the example given for the carbon dioxide molecule in Chapter 11. The mass-weighted Cartesian coordinates $\tilde{U}_i = \sqrt{m_i}\, U_i$ are introduced,[1] and the Hessian matrix in mass-weighted Cartesian coordinates is diagonalized. We have performed this step by hand in the Demonstration of Chapter 11. The square root of each obtained eigenvalue λ_k of the mass-weighted Hessian matrix is related to the **vibrational frequency** corresponding vibrational frequency f_k:

$$f_k = \frac{1}{2\pi} \sqrt{\lambda_k} \quad (12.2)$$

Six frequencies (five for linear molecules) are zero, because they refer to translational and rotational motions of the molecule, that is, only $3N - 6$

[1] For convenience we use the same index i for the atomic masses. Thus, $m_1 = m_2 = m_3$ corresponds to atom $I = 1$.

degrees of freedom are associated with vibrational frequencies ($3N-5$ for linear molecules). In practice, the eigenvalues for translational and rotational motions are not exactly zero due to numerical inaccuracies of the computer program, and frequencies of up to $10\,\text{cm}^{-1}$ might be observed, depending on the numerical accuracy of the computer implementation.

If two or more vibrational modes have the same frequency, they are called degenerate, and degeneracy is related to the symmetry of a **degeneracy** molecule (see the cyclobutadiene example in Chapter 10) and requires at least one rotational axis $C_n\,(n>2)$. The vibrational frequencies can be used to distinguish between a minimum on the potential energy surface and a transition state. In both cases the first derivatives of the potential with respect to the Cartesian coordinates are all zero. For a minimum of the potential energy surface, all eigenvalues must be positive, and hence the vibrational frequencies have positive real values. For a transition state, at least one eigenvalue of the mass-weighted Hessian matrix is negative and the corresponding frequency has an imaginary value due to the square root in Equation 12.2.[2]

12.2.2
Normal Modes

Each eigenvalue λ_k is associated with an eigenvector $\boldsymbol{q}_k$, which we here call, having $3N$ components q_{ik}. The eigenvectors can be used to construct a new set of coordinates Q_k from the mass-weighted Cartesian coordinates $\tilde{U}_i$. The new coordinates Q_k are called the **normal mode** normal modes and are defined as

$$\boldsymbol{Q}=\begin{pmatrix} Q_1 \\ Q_2 \\ \vdots \\ Q_{3N} \end{pmatrix}=\begin{pmatrix} \sum\limits_{i=1}^{3N} q_{i1}\,\tilde{U}_i \\ \sum\limits_{i=1}^{3N} q_{i2}\,\tilde{U}_i \\ \vdots \\ \sum\limits_{i=1}^{3N} q_{i3N}\,\tilde{U}_i \end{pmatrix}=\begin{pmatrix} \sum\limits_{i=1}^{3N} q_{i1}\sqrt{m_i}\,U_i \\ \sum\limits_{i=1}^{3N} q_{i2}\sqrt{m_i}\,U_i \\ \vdots \\ \sum\limits_{i=1}^{3N} q_{i3N}\sqrt{m_i}\,U_i \end{pmatrix}$$

(12.3)

---

2) Some computer codes quote imaginary frequencies with a negative sign. Remember that a vibrational frequency can never be negative: it is a real, positive value for minima, and an imaginary number (as not observable in nature) for a mode characterizing a maximum.

This very closely resembles the construction of molecular orbitals in an LCAO approach, with the molecular orbital coefficients as a multiplier matrix. The coefficients q_{ik} of a normal mode Q_k describe the displacement of each Cartesian coordinate U_i from the equilibrium position. The classification and naming of vibrations, for example, as bending or stretching mode, is an interpretation of the eigenvectors and refers to the most dominant movement of the atoms in that mode. For large molecules, the normal modes might get very complicated, and it is not always possible to label them accordingly. The computer animation of a molecular vibration plots the coordinates of the molecule in its equilibrium position plus the displacement $A\sin(2\pi f_k t)q_{ik}/\sqrt{m_i}$ for each Cartesian coordinate U_i. We see the molecule's motion associated with the eigenvalue λ_k (which corresponds to the mode's frequency f_k) on the screen and A is an arbitrary scaling factor to allow a reasonable visualization.

The normal modes are obtained by diagonalization, and they are therefore orthogonal to each other. This greatly simplifies the treatment of the vibrational modes of polyatomic molecules. Following Equation 12.3, we can transform the coordinates of the molecular system in such a way that we use the normal modes Q_k as coordinates. **vibrational Hamiltonian** With this transformation, the vibrational Hamiltonian, which usually depends on the Cartesian coordinates, can now be written in terms of the normal modes:

$$\hat{H}_{\mathrm{vib}} = \frac{1}{2}\sum_{i=1}^{N_{\mathrm{vib}}}\frac{\partial^2}{\partial Q_i^2} + \frac{1}{2}\sum_{i=1}^{N_{\mathrm{vib}}}\lambda_i Q_i^2 = \frac{1}{2}\sum_{i=1}^{N_{\mathrm{vib}}}\left(\frac{\partial^2}{\partial Q_i^2} + \lambda_i Q_i^2\right) \quad (12.4)$$

Here N_{vib} is the number of vibrational modes for the molecule, and the mass is present through the definition of Q (Equation 12.3). According to Equation 12.4, we can simplify the treatment of the vibrational system by solving N_{vib} independent harmonic oscillators, each of them in the same way as in Chapter 4. The vibrational energy is then conveniently calculated as the sum of the individual energies of the N_{vib} independent harmonic oscillators.

12.2.3
Symmetry in Normal Modes

The normal modes of the vibrations of a symmetric molecule adopt the same symmetry as the molecule itself. A thorough discussion of this **representation theory** topic requires representation theory for groups [1, 2]. Therefore, we will give only a brief introduction to the subject at this point, illustrated for the water molecule. The water molecule, which has C_{2v} symmetry, exhibits three vibrations: the symmetric stretching mode, the antisymmetric stretching mode, and the bending mode (Figure 12.1).

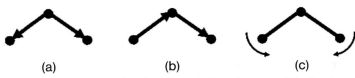

(a)   (b)   (c)

Figure 12.1 The vibrational modes of water (schematic representation): (a) symmetric stretching, (b) antisymmetric stretching, and (c) bending mode.

All the symmetry operations $(E, C_2, \sigma_1, \sigma_2)$ of the point group C_{2v} leave the symmetric stretching mode and the bending mode unchanged. For the antisymmetric stretching mode, the situation is different. The mirror plane σ_1 containing all three atoms and the symmetry element E leave the displacements of the atoms unchanged (lengthening of one bond and shortening of the other). In contrast, the rotational axis C_2 and the mirror plane σ_2, which is perpendicular to the molecular plane, interchange the lengthening and shortening of the two bonds. This behavior is summarized in Table 12.1 and Figure 12.2.

symmetry operation

A rigorous group theoretical treatment shows that, for the point group C_{2v}, there are not just two patterns. In fact, four different patterns exist, which are shown in Table 12.2. This is the character table of the point group C_{2v}, and each line is called an irreducible representation of C_{2v}. Character tables exist for all point groups [3]. It can be shown that normal modes always behave like an irreducible representation of the corresponding point group of the molecule. If normal modes are degenerate, together they form an irreducible representation, and the degeneracy is equal to the dimensionality of that representation.

character table
irreducible representation

Table 12.1 The vibrational modes of the water molecule (C_{2v}) transform differently upon application of different symmetry operations. A "1" denotes that a vibrational mode is not changed by a particular symmetry operation, and "-1" denotes that a mode is changed.

Mode	E	C_2	σ_1	σ_2
Symmetric stretching, bending	1	1	1	1
Antisymmetric stretching	1	-1	1	-1

Table 12.2 Full character table of the point group C_{2v}.

Irreducible representation	E	C_2	σ_1	σ_2
A_1	1	1	1	1
A_2	1	1	-1	-1
B_1	1	-1	1	-1
B_2	1	-1	-1	1

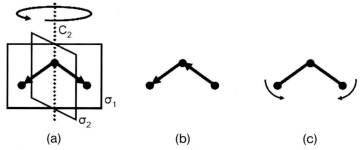

(a)	(b)	(c)

Figure 12.2 Results of different symmetry operations of the elements of the point group C_{2v} on the vibrational modes of the water molecule. (a) The symmetry elements of the water molecule. The symmetric stretching mode remains unchanged upon application of all four symmetry operations. (b) The symmetry elements C_2 and σ_2 interchange the antisymmetric mode. (c) The bending mode remains unchanged upon application of all four symmetry operations.

The irreducible representations are labeled according to the following rules. Here, we will discuss the first two only – the others are mentioned for completeness.

- A: no degeneracy; symmetric (1) upon rotation around the principal rotational axis.
- B: no degeneracy; antisymmetric (−1) upon rotation around the principal rotational axis.
- E: irreducible representation for two-fold degeneracy.
- T: irreducible representation for three-fold degeneracy.
- Subscript g:[3] symmetric (1) upon inversion.
- Subscript u:[3] antisymmetric (−1) upon inversion.

totally symmetric representation

The irreducible representation that leaves every vibration unchanged for all symmetry operations is called the totally symmetric representation. The normal modes of the symmetric stretching vibration and the bending vibration in the water molecule transform like the totally symmetric representation A_1 of the point group C_{2v}. The antisymmetric stretching mode transforms like the irreducible representation B_1. For linear molecules, Greek letters are commonly used to label irreducible representations. The vibrations in carbon dioxide ($D_{\infty h}$) transform like Σ_g^+ (symmetric stretching mode), Σ_u^+ (antisymmetric stretching mode), and Π_u (degenerate bending modes). Again, for a complete discussion of this topic, we refer the interested reader to the literature [1, 2].

---

3) The indices g and u are originated in the German words "gerade" and "ungerade" (Engl. even and odd).

12.2.4
Selection Rules

The intensity of transitions in spectroscopy is ruled by terms like **intensity**

$$\int\int\int \Psi_1(r)\,\hat{O}\Psi_2(r)\,\mathrm{d}x\,\mathrm{d}y\,\mathrm{d}z \tag{12.5}$$

where Ψ_1 is the wavefunction of the lower state, Ψ_2 is the wavefunction of the upper state, and $\hat{O}$ is the excitation operator (for Raman spectroscopy, the polarizability; for infrared spectroscopy, the dipole operator). The integral is carried out over the entire space. Only if the term $\Psi_1(r)\,\hat{O}\Psi_2(r)$ behaves like the totally symmetric representation of the corresponding point group can the value for the integral become non-zero. This can be easily rationalized if we remember that integrals of the type

$$\int_{-a}^{a} f(x)\,\mathrm{d}x \tag{12.6}$$

are zero for any antisymmetric function $f(x)$: for each point $(x, f(x))$ there is a counterpart $(-x, f(-x))$. For antisymmetric functions, the relation $f(x) = -f(-x)$ holds, and the sum of both points contributes zero to the integral.

The symmetry property of the term $\Psi_1(r)\,\hat{O}\Psi_2(r)$ is determined by the symmetry properties of each contributor, Ψ_1, $\hat{O}$, and Ψ_2. The irreducible representation of the product $\Psi_1(r)\,\hat{O}\Psi_2(r)$ is given by the multiplication of the irreducible representations for Ψ_1, $\hat{O}$, and Ψ_2. This is also called the direct product. As an example of a direct **direct product** product, we take two functions Ψ_1 and Ψ_2 that belong to the irreducible representations B_1 and B_2, respectively, of the point group C_{2v}. In order to determine the irreducible representation for the product function $\Psi_1(r) \cdot \Psi_2(r)$, we need to determine the direct product $B_1 \otimes B_2$. Multiplication of the irreducible representations for B_1 and B_2 shows that the product $\Psi_1(r) \cdot \Psi_2(r)$ belongs to the irreducible representation A_2 (see Table 12.3).

Table 12.3 Example of the direct product of two wavefunctions belonging to the irreducible representations B_1 and B_2 for the point group C_{2v}.

Function	Irreducible representation	E	C_2	σ_1	σ_2
Ψ_1	B_1	1	-1	1	-1
Ψ_2	B_2	1	-1	-1	1
$\Psi_1 \cdot \Psi_2$	$B_1 \otimes B_2 = A_2$	$1\times1=1$	$(-1)\times(-1)=1$	$1\times(-1)=-1$	$(-1)\times1=-1$

Table 12.4 Complete character table of the point group C_{2v}.

C_{2v}	E	$C_2(z)$	$\sigma_v(xz)$	$\sigma_v(yz)$	Linear functions, rotations	Quadratic functions	Cubic functions
A_1	$+1$	$+1$	$+1$	$+1$	z	x^2, y^2, z^2	z^3, x^2z, y^2z
A_2	$+1$	$+1$	-1	-1	R_z	xy	xyz
B_1	$+1$	-1	$+1$	-1	x, R_y	xz	xz^2, x^3, xy^2
B_2	$+1$	-1	-1	$+1$	y, R_x	yz	yz^2, y^3, x^2y

If both functions Ψ_1 and Ψ_2 belong to the same irreducible representation, the product $\Psi_1 \cdot \Psi_2$ obviously always transforms like the totally symmetrical irreducible representation, since multiplications yield $(-1)^2 = 1^2 = 1$.

In order to determine the irreducible representation for the term $\Psi_1(\boldsymbol{r}) \, \hat{O} \, \Psi_2(\boldsymbol{r})$, we also need to know the irreducible representation for the excitation operator $\hat{O}$. For IR spectroscopy, the excitation **dipole moment operator** operator is given by the dipole moment operator; for Raman spectros- **polarizability operator** copy, it is given by the polarizability operator. Instead of working with the full operator, it is sufficient to know that the dipole moment operator transforms like the Cartesian coordinates x, y, z of the corresponding point group, whereas the polarizability operator transforms like the quadratic coordinates $(x^2, y^2, z^2, xy, xz, yz)$. The irreducible representations for the Cartesian coordinates and the quadratic coordinates can be found in the last columns of most character tables. The character tables on the CD show the irreducible representations for the rotations (R_x, R_y, R_z) and the cubic coordinates $(x^3, y^3, z^3, x^2y, x^2z, xy^2, y^2z, xz^2, yz^2, xyz)$ as well. Table 12.4 now shows the complete character table for the point group C_{2v}.

Now we have to determine the irreducible representations for the ground-state function Ψ_1 and the excited-state function Ψ_2. From Chapter 4 (see Figure 4.1) we know that the vibrational wavefunction $\Psi(Q_k)$ for even quantum numbers n is always symmetric with respect to an inversion of the coordinate Q_k. This means that for even quantum numbers n the vibrational wavefunction $\Psi(Q_k)$ always transforms like the totally symmetric irreducible representation for any normal mode. For odd quantum numbers n the vibrational wavefunction $\Psi(Q_k)$ is antisymmetric with respect to an inversion of the coordinate Q_k, and $\Psi(Q_k)$ transforms like the corresponding normal mode Q_k. Figure 12.3 shows the irreducible representations for several vibrational states of a B_1 normal mode of the point group C_{2v}.

For a vibrational excitation from $n = 0$ to $n = 1$ (the fundamental excitation), the ground-state wavefunction Ψ_1 belongs to the totally

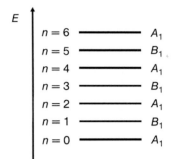

Figure 12.3 Vibrational states for a B_1 normal mode of the point group C_{2v}.

symmetric irreducible representation, and the symmetry of the term $\Psi_1\,\hat{O}\,\Psi_2$ is determined by the symmetry of the term $\hat{O}\,\Psi_2$ only. As discussed above, the excitation operator $\hat{O}$ and the excited-state wavefunction Ψ_2 must belong to the same irreducible representation in order to form a totally symmetric product $\hat{O}\Psi_2$. Keeping in mind how the excitation operators for infrared spectroscopy and Raman spectroscopy transform, we finally arrive at the general selection rules **selection rule** for infrared and Raman spectroscopy.

- The only molecular vibrations (normal modes) that are infrared- **IR** active are those that transform as an irreducible representation of one of the Cartesian axes (x, y, z; dipole moment operator).

- The only molecular vibrations (normal modes) that are Raman-active **Raman** are those that transform as an irreducible representation of one of the products of two Cartesian axes (x^2, y^2, z^2, xy, xz, yz; polarizability operator).

The solutions of the Schrödinger equation (see Chapter 4 and Equation 4.10) are $E_n = (n + \frac{1}{2})\hbar\omega_0$ (with $n = 0, 1, 2, \ldots$), resulting in equally spaced vibrational levels with a spacing of $\hbar\omega_0$. Within the harmonic approximation, only vibrational excitations with $\Delta n = 1$ are allowed. At room temperature, only transitions from the ground state ($n = 0$) to the first level ($n = 1$) can be observed. At higher temperatures, excited vibrational levels are also populated, and transitions from $n = 1$ or even higher can be observed (so-called hot bands). Overtones ($\Delta n > 1$) or combination tones (simultaneous excitation of more than one vibration) are not allowed within the harmonic approximation and require a more sophisticated treatment of the vibrations. Their treatment is beyond the scope of this book, and the interested reader is referred to the literature [4].

12.3
Demonstration

We use the molecule 1,4-difluorobenzene $C_6H_4F_2$ as an example for the analysis of a polyatomic molecule. The molecule consists of 12 atoms, leading to 30 degrees of freedom for vibrational motions. Since the molecule has no rotational axis C_n ($n > 2$), we do not expect degenerate vibrational states, and all 30 normal modes should belong to different vibrational frequencies.

We start with the determination of the irreducible representations for the normal modes [3]. The point group for 1,4-difluorobenzene is D_{2h}, and the 30 normal modes belong to the following irreducible representations:

$$6\ A_g, 5\ B_{1g}, 3\ B_{2g}, 1\ B_{3g}, 2\ A_u, 3\ B_{1u}, 5\ B_{2u}, 5\ B_{3u}$$

The Cartesian axes x, y, z transform like B_{1u}, B_{2u}, B_{3u}, and only these 13 normal modes can be observed in IR spectroscopy. The quadratic Cartesian coordinates x^2, y^2, z^2, xy, xz, yz transform like A_g, B_{1g}, B_{2g}, B_{3g}, and only these 15 normal modes can be observed in Raman spectroscopy. The two A_u normal modes cannot be observed, in neither infrared nor Raman spectroscopy.

For the determination of the vibrational frequencies and intensities in IR and Raman spectroscopy, we perform a calculation with modern computational chemistry software, for example, the program deMon delivered with this book. The calculation may take some time; therefore the output file and the Molden visualization file can be found on the CD in the folder examples. First, the equilibrium geometry of 1,4-difluorobenzene is determined by performing a geometry optimization (see Chapter 9). The number of independent internal coordinates using symmetry constraints is always equal to the number of totally symmetric vibrations. Thus, for the 1,4-difluorobenzene molecule in D_{2h} symmetry, only six internal coordinates are needed to fully describe the geometry (see Table 12.5).

Table 12.5 Bond lengths and angles of the optimized geometry of 1,4-difluoro-benzene. The superscripts on the carbon atoms denote their non-carbon neighbors.

Atoms	Distance (Å)	Atoms	Angle (deg)
$C^F - C^H$	1.40	$C^H - C^F - C^H$	122.2
$C^H - C^H$	1.41	$C^F - C^H - H$	119.7
$C^F - F$	1.37	$^a C^F - C^H - C^H$	118.9
$C^H - H$	1.09	$^a C^H - C^F - F$	118.9

^{a}These last two angles are redundant coordinates and not necessary to fully describe the molecule.

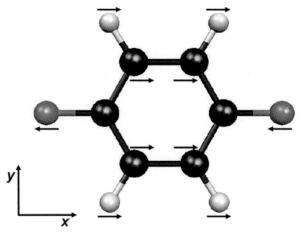

Figure 12.4 Schematic view of the normal mode Q_{10} of 1,4-difluorobenzene. The most dominant displacements from the atomic equilibrium positions are indicated by arrows.

For the optimized geometry, the Hessian matrix and the intensities are calculated. For a molecule of that size, the calculation may take several hours, so again the output file is provided on the CD. The irreducible representations are determined manually by inspection of the normal modes. This procedure is explained in detail for the normal mode Q_{10} as a typical example. The most dominant displacements from the atomic equilibrium positions are indicated in Figure 12.4 by arrows.

Obviously, the normal mode Q_{10} is antisymmetric with respect to inversion (the inversion center is in the center of the carbon ring) and with respect to reflection at the mirror plane $\sigma(yz)$. However, it is symmetric with respect to reflection at mirror planes $\sigma(xy)$ and $\sigma(xz)$. The irreducible representation B_{3u} fulfills these requirements. By crosschecking we see that Q_{10} is indeed symmetric with respect to rotation around $C_2(x)$, but antisymmetric with respect to rotations around $C_2(y)$ and $C_2(z)$. The calculated spectroscopic data including assignment to an irreducible representation for all normal modes is summarized in Table 12.6.

The irreducible representations are determined manually by inspection of the normal modes. For vibrations 4 and 15, the calculated IR and Raman intensities are close to zero. This is not surprising because these are A_u modes, which are neither infrared- nor Raman-active. Several other vibrations (nos. 9, 14, 22, 23) also show intensities close to zero, although they are not forbidden by symmetry. All vibrations can be visualized with the program Molden, which also uses the calculated data to draw IR and Raman spectra (see Figure 12.5).

Table 12.6 Calculated vibrational frequencies, IR and Raman intensities for 1,4-difluorobenzene. Columns 4 and 6 refer to the group theoretical treatment in the harmonic approximation.

No.	Frequency (cm^{-1})	Irreducible representation	IR-active	IR intensity $(km\ mol^{-1})$	Raman-active	Raman intensity $(Å^4\ amu^{-1})$
1	155.0	B_{1u}	yes	30.8	no	0.0
2	331.2	B_{2u}	yes	3.5	no	0.2
3	359.4	B_{2g}	no	0.2	yes	4.0
4	412.3	A_u	no	0.3	no	0.0
5	426.3	B_{1g}	no	0.0	yes	5.2
6	439.9	A_g	no	0.0	yes	13.4
7	523.2	B_{1u}	yes	66.8	no	0.0
8	624.0	B_{1g}	no	0.0	yes	59.5
9	681.2	B_{2g}	no	0.2	yes	0.0
10	717.1	B_{3u}	yes	38.4	no	0.3
11	764.4	B_{3g}	no	2.6	yes	2.3
12	802.2	B_{1u}	yes	94.0	no	0.1
13	841.1	A_g	no	0.0	yes	202.5
14	881.0	B_{2g}	no	0.5	yes	0.1
15	900.6	A_u	no	0.1	no	0.0
16	974.8	B_{3u}	yes	14.6	no	1.4
17	1078.6	B_{2u}	yes	10.9	no	1.5
18	1121.7	A_g	no	0.0	yes	19.9
19	1172.8	B_{3u}	yes	13.9	no	0.9
20	1248.6	B_{1g}	no	0.0	yes	15.7
21	1249.2	B_{1g}	no	0.0	yes	49.5
22	1361.4	B_{2u}	yes	0.0	no	0.0
23	1411.2	B_{2u}	yes	0.6	no	0.5
24	1494.8	B_{3u}	yes	117.5	no	1.3
25	1612.5	B_{1g}	no	0.0	yes	43.3
26	1618.7	A_g	no	0.0	yes	17.9
27	3147.0	A_g	no	0.0	yes	115.2
28	3151.4	B_{2u}	yes	0.5	no	2.8
29	3156.6	B_{3u}	yes	2.8	no	0.3
30	3178.9	A_g	no	0.0	yes	349.3

12.4
Problems

1. Irreducible representations of normal modes

Determine the irreducible representations for the normal modes in dichloromethane, ethane and ammonia. Decide for each mode whether it is infrared-active or Raman-active. If you are familiar with

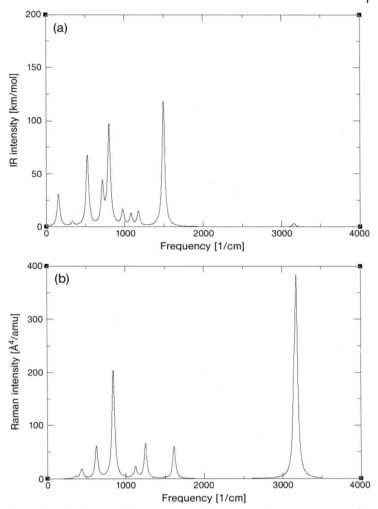

Figure 12.5 (a) Simulated IR spectrum and (b) simulated Raman spectrum of 1,4-difluorobenzene.

the group theoretical procedure, you are encouraged to perform the analysis by hand.

Note: You can carry out this task by using the program "Character Tables for Chemically Important Point Groups" on the CD.

2. Frequency analysis using computational chemistry software

Perform a geometry optimization and a frequency analysis of the molecules discussed in problem 1. Compare the calculated frequencies with experimental data from the literature [5].

Note: You may carry out these calculations with the `deMon` program.

3. Visualization of normal modes

Use the eigenvectors of the normal modes for dichloromethane, ethane and ammonia to visualize the vibrations. Sketch each vibration by drawing the molecular structure and attaching appropriately scaled arrows showing the direction of the displacement. Try to characterize each mode (e.g., as stretching or bending mode). Repeat the geometry optimization and frequency analysis as in problem 2. Visualize the molecular vibrations using `molden`.

4. Comparison of infrared and Raman spectra

For 1,2-dichloroethene, two isomers, the *cis-* and the *trans-* forms, exist. Determine the symmetry of the normal modes for both isomers. Discuss how both isomers can be distinguished by IR and Raman spectroscopy. Perform a geometry optimization and frequency analysis using computational chemistry software, discuss the results, and compare them with the literature [5].

5. Characterization using infrared and Raman spectroscopy

Determine the symmetry of the normal modes for 1,1-dichloroethene, and perform a geometry optimization and frequency analysis. Use the results from problem 4 and decide if it is possible to distinguish between these three isomers by IR and Raman spectroscopy.

12.5
Review and Summary

In this chapter, we have gained a deeper insight into the vibrational properties of molecules. We have diagonalized the $3N \times 3N$ Hessian matrix – this time using a single-vector representation of the $3N$ mass-weighted Cartesian coordinates of the molecule – to obtain its eigenvalues, the vibrational frequencies f_k. There are $3N - 6$ degrees of freedom ($3N - 5$ for linear molecules) that contribute to the vibrational frequencies. The remaining frequencies are zero (due to numerical inaccuracies they might have values of up to $10\,\mathrm{cm}^{-1}$), because they refer to translational and rotational motions of the molecule. Vibrations of the same frequency (and energy) are called degenerate. If the molecule is in a transition state, at least one eigenvalue of the mass-weighted Hessian matrix is negative and has a corresponding imaginary frequency has an imaginary value.

To each frequency (eigenvalue) there exists a normal mode Q_k (eigenvector). The normal mode vector describes the direction and the relative strength of the displacement of each atom in a molecule for a specific vibration k. The naming of vibrations refers to the most dominant movement of the atoms in the mode.

The treatment of the vibrational system is simplified to a great extent by solving independent harmonic oscillators, and the vibrational energy is calculated as the sum of the individual energies of the independent harmonic oscillators in a new basis.

Furthermore, we have discussed the symmetry of molecular vibrations, introducing group theory and character tables. Most importantly, we learned that vibrations may be symmetric or antisymmetric with respect to a certain symmetry operation. This information is summarized in sets of irreducible representations in the character tables. Character tables are specific for every point group. Finally, we have discussed briefly the infrared and Raman selection rules, which determine the intensities of the molecular vibrations. For IR excitations, the dipole operator determines the intensity; for Raman excitations, it is the polarizability operator.

References

1 Cotton, F.A. (1990) *Chemical Applications of Group Theory*, Wiley-Interscience.

2 Bishop, D.M. (1993) *Group Theory and Chemistry*, Dover Publications.

3 A collection of character tables for all the chemically important point groups can be found on the accompanying CD or at the web page of Jacobs University, Bremen (http://symmetry.jacobs-university.de).

4 Califano, S. (1976) *Vibrational States*, Wiley-Interscience.

5 An excellent source of spectroscopic data can be found at the home page of the National Institute of Standards of the USA (http://webbook.nist.gov/chemistry).

13
Ionization Potential and Electron Affinities of Molecules

13.1
Aim

We calculate the ionization potential and electron affinity of molecules. In contrast to atoms, experiments on molecules show different values for the ionization potential and electron affinity, depending on the time scale of the measurement, and depending on the initial conditions. We will understand the reason for this behavior. In fast processes, the geometry of atom and ion remains approximately the same (Franck–Condon principle); while in slow processes, the energy is measured between the equilibrium geometry of the neutral and ionic forms of the molecule (adiabatic case). We investigate small, simple molecules (hydrogen H_2 and hydrogen fluoride HF) and a recently observed all-metallic ion, Al_4^{2-}.

13.2
Theoretical Background

13.2.1
Field of Application

The theoretical background of the ionization potential (IP) and the electron affinity (EA) has been discussed in Chapter 6 and will not be repeated here. We extend the calculations of IPs and EAs to molecules, as these are characteristic molecular quantities. Experimentally, they can be accessed via laser spectroscopy. These spectroscopic experiments are particularly important if the substance of interest is only available in the gas phase, for example, in a mass spectrometer. As the time of measurement is very short in laser spectroscopy, molecules and in particular molecular ions of a short lifetime can be investigated in this way. Comparison of calculated and measured IPs and EAs allow the

ionization potential
electron affinity

Computational Chemistry Workbook: Learning Through Examples
Thomas Heine, Jan-Ole Joswig, and Achim Gelessus
Copyright © 2009 WILEY-VCH Verlag GmbH & Co. KGaA, Weinheim
ISBN: 978-3-527-32442-2

identification of the structure of a molecular fragment that has a short lifetime and is present only in an ion beam in a vacuum chamber.

13.2.2
Influence of Geometry

A change of the electronic structure of a molecule changes its bonding properties and results in some rearrangement of its structure. Usually, the new equilibrium structure has the same topology (that is, the same bonding pattern) as the neutral molecule, but quantitative changes of the geometry are observed. Such rearrangements take place if electrons are transferred to or from the molecule, that is, for reactions

$$A \rightarrow A^+ + e^- \quad \text{or} \quad A + e^- \rightarrow A^-$$

and so on. The equilibrium structure of the ion is different from the equilibrium structure of the neutral molecule.

It is now important to understand the experimental situation. Owing to the large difference of masses, electron transitions are fast compared to the relaxation of the molecular structure. Therefore, we can distinguish two extreme cases.

If the measurement process takes a time that is in the order of the molecular relaxation (about 10^{-13} s $= 100$ fs) or longer, we measure the energy difference of neutral and charged molecules, each time in their equilibrium positions. These quantities are called the adiabatic ionization potential (IP_{ad}) and the adiabatic electron affinity (EA_{ad}), respectively.

adiabatic ionization potential
adiabatic electron affinity

In contrast, if the measurement process takes a short time, close to the time of an electron transition from a bound state to vacuum or vice versa (about 10^{-15} s $= 1$ fs), the ionic and neutral molecule have – within a good approximation – the same structure, namely the structure of the system before the electron transfer was started. This corresponds to the Franck–Condon principle. In this case we speak of the vertical ionization potentials (IP_v) and the vertical electron affinity (EA_v). Both the vertical and the adiabatic IPs are illustrated in Figure 13.1a, where the potential energy surfaces of the neutral particle (low-energy curve) and of the ionized particle (high-energy curve) are shown.

Franck–Condon principle
vertical ionization potential
vertical electron affinity

The two curves in Figure 13.1a have their minima at different positions corresponding to different equilibrium geometries. Before the experiment, the structure is at the equilibrium position of the low-energy curve. In a short-time experiment, the measured ionization potential is the vertical IP, and corresponds to the difference between the two potential energy surfaces at the equilibrium geometry of the neutral species. If the measurement is on a long time scale, the ionized molecule can relax, and moves to the minimum of the high-energy

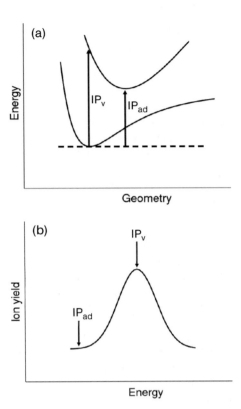

Figure 13.1 (a) Illustration of the vertical ionization potential (IP_v) and the adiabatic ionization potential (IP_{ad}). The lower curve corresponds to the potential energy surface of the neutral molecule, and the upper curve to that of the charged species. (b) Ion yield in a photoelectron spectroscopy experiment. The left edge of the curve is determined by the adiabatic IP, while the maximum is given by the vertical IP.

curve. Consequently, the adiabatic IP is the energy difference between the two minima. Obviously, adiabatic IPs must be lower than vertical IPs, as the energy difference is always reduced in the relaxation of the ion. A typical result from experiment is shown in Figure 13.1b. The IP signal appears with a low-energy flank of the ion yield; this position indicates where first ionizations are possible and therefore corresponds to the adiabatic IP – the lowest possible ionization potential. At increasing energy, the ion yield increases until it reaches a maximum. The maximum corresponds to the vertical ionization potential, and refers to the highest probability that photons can ionize the neutral particle. Afterwards, the ion yield reduces again. There is still an ion yield, as ionizations occur, for example, vertically, starting from some non-equilibrium geometry of the neutral molecule. In summary, the left flank of the ion yield curve corresponds to the adiabatic IP, while the maximum corresponds to the vertical IP.

The discussion of vertical and adiabatic IPs can be transferred directly to the EA. Here, we can observe two vertical values, the vertical EA, which corresponds to the transfer of one electron from vacuum to the neutral molecule, and the vertical detachment energy (VDE), where an electron is emitted from the anion to vacuum, and a neutral particle is formed (Figure 13.2).

vertical detachment energy

All values can be calculated in the same way as discussed for atoms in Chapter 6. This holds for the energy differences between charged and neutral molecules. It also holds for the vertical ionization potential within the Koopmans' theorem – though this is a rough approximation also for molecules.

13.3
Demonstration

We use the oxygen molecule O_2 for the demonstration. We first optimize the neutral molecule – note that it has a multiplicity of $M = 3$ (see Chapter 10) – and save the equilibrium structure. If you work with deMon (see Technical Details), you will find an equilibrium distance of 1.236 Å. The HOMO energy, which is the energy of the highest spin-up orbital, is $\varepsilon_{HOMO} = -0.25466$ Hartree,

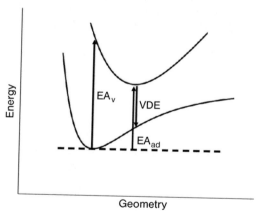

Figure 13.2 Illustration of the vertical electron affinity (EA_v), the adiabatic electron affinity (EA_{ad}), and the vertical detachment energy (VDE). The lower curve corresponds to the potential energy surface of the neutral molecule, and the upper curve to that of the charged species.

and corresponds, according to Koopmans' theorem, to a vertical ionization potential of $IP_v^{(Koopmans)} = 6.93$ eV. Experiment reports an IP value of 12.07 eV, which is significantly off the prediction following Koopmans' theorem. The total energy of the optimized, neutral O_2 molecule is -150.2127 Hartree.

Now we calculate the energy of O_2^+, the cation of oxygen, which has a multiplicity of $M = 2$, at the equilibrium geometry of the neutral species, that is, we only change the multiplicity. The calculation yields a total energy of -149.7440 Hartree. This results in a vertical ionization potential of $IP_v = 12.75$ eV, in very reasonable agreement with experiment. Optimization of the cation lowers the O_2 bond length to 1.141 Å. This reduction of the bond length can be expected, because the MO diagram tells us that an antibonding electron has been removed. The total energy of the cation in its equilibrium structure has a value of -149.7585 Hartree, which results in a slightly lower adiabatic ionization energy of $IP_{ad} = 12.36$ eV.

The vertical electron affinity is assessed by the difference of the total energies between the neutral and anionic forms of O_2 (both in the neutral geometry). For O_2^-, we have to set the multiplicity again to $M = 2$, and we obtain a total energy of -150.1938 Hartree, giving a value for the vertical EA of $EA_v = 0.51$ eV. This predicts an unstable species. To calculate the adiabatic EA, we optimize O_2^- and find a stretched bond length of 1.3840 Å – another electron has filled one of the antibonding orbitals – and a total energy of -150.2136 Hartree. Indeed, we find that the anion is stabilized: the energy of the optimized anion is slightly lower than the energy of the neutral particle. The adiabatic electron affinity is now given as 0.02 eV. Finally, we compute the vertical detachment energy by calculating the neutral oxygen molecule at the geometry of O_2^-. We obtain a total energy of -150.1927 Hartree, resulting in a value for the VDE $= 0.57$ eV.

13.4
Problems

1. Ozone

Calculate IP_v, IP_{ad}, EA_v, EA_{ad} and VDE for ozone. Compare your results with the literature, for example, at http://webbook.nist.gov/chemistry.

2. Planar Al_4^{n-} clusters

Recently, the formation of planar all-metal clusters Al_4^{2-} [1] and Al_4^{4-} [2] has been reported in the literature. The clusters have been characterized by VDE measurements of $Li_xAl_4^{n-}$ species. Calculate the VDE of Al_4^{2-}.

3. Adiabatic versus vertical IP and EA – dependence on system size

The differences between adiabatic and vertical IPs and EAs depend on the size of the molecule except in those cases where the HOMO is a strongly localized orbital. Investigate the differences in adiabatic and vertical IPs for a series of molecules with increasing size: hydrogen (H_2), methane (CH_4), benzene, and naphthalene.

13.4.1
Technical Details

If you use deMon, your input should be as follows (note the improved auxiliary basis requested by the keyword AUXIS (GEN-A2*)):

```
TITLE IP AND EA
VXCTYPE AUXIS PBE
BASIS (DZVP-GGA)
AUXIS (GEN-A2*)
CHARGE 0
GEOMETRY CARTESIAN
...
END
```

For the calculation of anion or cation, you only need to change the charge to −1 or + 1, respectively. The multiplicity is taken automatically to the lowest possible value, but can also specified in the input as discussed in Chapter 10. Intermediate structures can be obtained either by manipulating the .new output, by copying and pasting from the deMon output, or by saving optimized structures directly in Molden.

For the O_2 molecule (MULTIPLICITY 3), orbital energies of spin-up (ALPHA MO ENERGIES) and spin-down electrons (BETA MO ENERGIES) are calculated and stated separately in the output file. Use the highest-energy orbital that is occupied as HOMO.

13.5
Review and Summary

You have learned that the theoretical principles for the calculation of electron affinities and ionization potentials is the same for atoms and for molecules. As the number of electrons is different in neutral and charged species, the chemical bonding changes too, and the equilibrium structures are different. Owing to this difference, the ionization potential and electron affinity depend on the time scale of the experiment. For short time scales, the Franck–Condon principle applies, and we can calculate the values for IP and EA at the geometry of the initial structures. To indicate that we compare to energies at the same geometry (see Figure 13.1), we speak of vertical values: vertical ionization potential (IP_v), vertical electron affinity (EA_v), and vertical detachment energy (VDE). In contrast, measurements on long time scales allow the molecule to relax its geometry, and adiabatic values are measured: adiabatic ionization potential (IP_{ad}) and adiabatic electron affinity (EA_{ad}).

We learned that calculating the IP through Koopmans' theorem does not always give satisfactory results for molecules, as the electronic structure relaxes after losing one electron. The approximation should not be used for serious calculations, as its computational advantage is rather small.

Computations of electron affinities, as all calculations involving negatively charged species, should be carried out with care. As more electrons are in the system, the spatial extension of the electron density increases, and large, comfortable basis sets should be used.

References

1 Li, X., Kuznetsov, A.E., Zhang, H.F., Boldyrev, A.I. and Wang, L.S. (2001) *Science*, **291**, 859.

2 Kuznetsov, A.E., Birch, K.A., Boldyrev, A.I., Li, X., Zhai, A.I. and Wang, L.S. (2003) *Science*, **300**, 622.

14
Thermochemistry

14.1
Aim

In this experiment we will connect quantum chemical calculations and statistical thermodynamics. We will then be able to calculate temperature-dependent properties, such as enthalpies and entropies, as well as all derived properties like heat capacities or free enthalpies. Through this approach, the energetics of chemical reactions and the temperature dependence of the distribution of isomers can be investigated.

14.2
Theoretical Background

If one solves the Schrödinger equation by performing a quantum chemical calculation for a molecule, the calculation results in the potential energy, which is often named the total energy, as it refers to the total energy of the electrons in the potential of the frozen nuclei. Minimizing the total energy by varying the atomic positions (the atomic coordinates) is the basic procedure of a geometry optimization (see Chapter 9). Thereby, the potential energy of the molecule is used as a measure of the stability of the system. For example, by comparing the total energies of two structural isomers, one can identify the more stable one. Another example for the usage of the property "potential energy" is to calculate reaction energies using the potential energies of the products and reactants (educts) of a reaction.

The question we start to answer in this chapter is this: How can we connect the potential energy of a single molecule, alone and at rest in a hypothetical universe, to quantities that can be measured in reality? We have already learned (Chapter 4) that the potential energy needs to be corrected by the zero-point energy of the nuclei. As we know from fundamental physical chemistry, we distinguish between energy and

Computational Chemistry Workbook: Learning Through Examples
Thomas Heine, Jan-Ole Joswig, and Achim Gelessus
Copyright © 2009 WILEY-VCH Verlag GmbH & Co. KGaA, Weinheim
ISBN: 978-3-527-32442-2

enthalpy, and for thermodynamic considerations it is important to calculate the free energy and the free enthalpy, which are functions of the system's temperature. We report them in the unit $kJ\,mol^{-1}$ or $kcal\,mol^{-1}$, and in the laboratory we mean the unit literally, per mole, per 6.022×10^{23} particles. Here, we will work on the connection between the energy of a single molecule, and that of an ensemble of particles. There are more factors that affect the energy of a macroscopic system, most importantly the intermolecular interactions. If the investigated molecules are in the gas phase, we assume that the intermolecular interactions are negligible. The situation can be very different in other situations, for example, in liquid or solid phase or in solution. The treatment of such effects is possible, but will not be discussed here.

Even if we want to describe molecules in the gas phase, the potential energy, calculated in a geometry optimization, is not equal to the internal energy of the system. First, as the nuclei are not moving, the potential energy of a molecule is valid only for a single and classically treated particle *in vacuo* at $T = 0$ K. Therefore, there is no direct relation to the macroscopic world in terms of measurable quantities, such as reaction energies. Moreover, as we learned in Chapter 4, the zero-point energy contributes to the energy of a molecule, even at $T = 0$. We show here that we can calculate energies that can be compared to experimental heats of reactions by treating the degrees of freedom of the molecule. For this purpose, we calculate the thermodynamic properties of molecules by combining quantum mechanics with statistical thermodynamics. The quantum chemical calculations result in the necessary parameters: the potential energy and several properties that are needed to calculated the partition function.

statistical
thermodynamics

14.2.1
Calculating Thermodynamic Functions Using the Partition Function

In the following, we are going to discuss an approximate way of calculating the internal energy U, the entropy S, the free energy F, the enthalpy H, the free enthalpy G, and the heat capacity C_V. It is possible to calculate all these thermodynamic properties as functions of the potential energy E_{pot} of the molecule, the partition function Q, and the vibrational energies E_i^{vib} (these are also needed to calculate Q). Moreover, the molecular mass and the tensor of the moments of inertia (or the rotational constants) are needed. You may find the series of equations below tedious and cumbersome, in particular if you imagine computing all those sums, products, logarithms, and so on by hand. It is therefore advisable to program the following expressions in a convenient way, for example, by a math processor, a spreadsheet, or, as on our CD, in a script language.

partition function

The internal energy at absolute zero $U(T=0\,\text{K})$ is calculated as the sum of the potential energy E_{pot} and the zero-point energy E_{ZPE}. As shown in Chapter 12 (Equation 12.4), all the oscillations of the molecule can be treated individually, and are hence a sum over N_{vib} individual vibrational contributions E_i^{vib}:

$$U(0) = U(T = 0\text{K}) = E_{\text{pot}} + E_{\text{ZPE}}$$

$$= E_{\text{pot}} + \frac{1}{2}\sum_{1}^{N_{\text{vib}}} E_i^{\text{vib}} = E_{\text{pot}} + \frac{1}{2}\sum_{1}^{N_{\text{vib}}} hf_i \qquad (14.1)$$

With the internal energy at absolute zero $U(0)$, we can write the internal energy U and the entropy S as functions of the temperature T [1]:

internal energy

entropy

$$U(T) = U(0) - \left(\frac{\partial \ln Q}{\partial \beta}\right)_V \qquad (14.2)$$

and

$$S(T) = \frac{U(T) - U(0)}{T} + k_B \ln Q \qquad (14.3)$$

In Equations 14.2 and 14.3 and in the following, the inverse temperature $\beta = 1/(k_B T)$ is used and k_B denotes Boltzmann's constant.

The free energy F is defined as

free energy

$$F(T) = U(T) - TS(T) \qquad (14.4)$$

Thus, for absolute zero, we find $F(0) = U(0)$. Using the terms for the internal energy and the entropy as defined above in Equations 14.2 and 14.3, we obtain

$$F(T) = U(0) - k_B T \ln Q = F(0) - k_B T \ln Q \qquad (14.5)$$

In the following, we will have to use the pressure p to calculate the enthalpy H and the free enthalpy G. We first derive the total derivative dF from Equation 14.4 by using the Gibbs equation $(dU = T\,dS - p\,dV)$:

$$dF = dU - T\,dS - S\,dT$$

$$= \underbrace{T\,dS - p\,dV}_{dU} - T\,dS - S\,dT = -p\,dV - S\,dT \qquad (14.6)$$

Since we examine an isothermal process $(dT = 0)$, it follows $dF = -p\,dV$. We can calculate the pressure using Equation 14.5 (and the fact that $F(0)$ is a constant):

$$p = -\left(\frac{\partial F}{\partial V}\right)_T = -\left(\frac{\partial}{\partial V}[F(0) - k_B T \ln Q]\right)_T = k_B T \left(\frac{\partial \ln Q}{\partial V}\right)_T$$

$$(14.7)$$

Finally, we can also calculate the enthalpy H, using $H = U + pV$ and Equations 14.2 and 14.7,

enthalpy

$$H(T) = U(T) + pV = \underbrace{U(0) - \left(\frac{\partial \ln Q}{\partial \beta}\right)_V}_{U(T)}$$

$$+ \underbrace{k_B T \left(\frac{\partial \ln Q}{\partial V}\right)_T}_{p} \times VH(0) - \left(\frac{\partial \ln Q}{\partial \beta}\right)_V + k_B TV \left(\frac{\partial \ln Q}{\partial V}\right)_T$$

$$(14.8)$$

free enthalpy and the free enthalpy G, using $G = F + pV$ and Equations 14.4 to 14.7,

$$G(T) = G(0) - k_B T \ln Q + k_B TV \left(\frac{\partial \ln Q}{\partial V}\right)_T \tag{14.9}$$

The equations above are formally exact. However, the partition function is not yet directly accessible. Using the ideal-gas approximation we can circumvent this problem and calculate the partition function through molecular contributions [2]. Moreover, the ideal-gas approximation also simplifies other relations.

14.2.2
Thermochemistry Within the Ideal-Gas Approximation

ideal-gas The ideal-gas approximation allows us simply to add up the internal
approximation energies of an ensemble of particles (molecules), because the intermolecular interactions are neglected in this approximation. Here, we can also understand the conversion of molecular units (Hartree, Rydberg, eV) into macroscopic energy units (kJ mol^{-1} or kcal mol^{-1}). Moreover, the relation between internal energy and enthalpy is simplified through the ideal-gas equation $pV = nRT$:

$$H = U + pV \Rightarrow H(0) = U(0) + nRT = U(0) + Nk_B T$$

$$(14.10)$$

and

$$H(T) = U(T) + nRT = U(T) + Nk_B T \tag{14.11}$$

The left-hand side of Equation 14.10 – the original thermodynamic definition $H = U + pV$ – is a macroscopic equation that is scaled down to a single-molecule property by the term nRT or $Nk_B T$. These terms are equal as long as the correct units are used consequently.

For an ideal gas we can use the following substitution (cf. Equation 14.7):

$$pV = k_B T \left(\frac{\partial \ln Q}{\partial V}\right)_T \times V = nRT \tag{14.12}$$

For indistinguishable molecules the partition functions of the ensemble, Q, and the partition function of the individual molecule, q, are connected in a simple way through $Q = q^N/N!$. Using the Stirling formula,[1] we end up with an expression for the internal energy U,

$$U(T) = U(0) - \left(\frac{\partial \ln Q}{\partial \beta}\right)$$

$$= U(0) + Nk_B T^2 \frac{\partial \ln q}{\partial T}$$

$$= E_{pot} + E_{ZPE} + Nk_B T^2 \frac{\partial \ln q}{\partial T} \qquad (14.13)$$

and the entropy S,

$$S(T) = \frac{U(T) - U(0)}{T} + k_B \ln Q$$

$$= Nk_B T \frac{\partial \ln q}{\partial T} + k_B (N\ln q - N\ln N + N) \qquad (14.14)$$

for a single molecule, $N = 1$ and $\ln N = 0$. Equation 14.14 simplifies[2] to

$$S(T) = Nk_B \left(T \frac{\partial \ln q}{\partial T} + \ln q + 1\right). \qquad (14.15)$$

With Equations 14.12 to 14.15 we can now calculate expressions for H, F, and G using the Gibbs equations:

$$H(T) = U(T) + pV$$

$$= E_{pot} + E_{ZPE} + Nk_B T^2 \frac{\partial \ln q}{\partial T} + Nk_B T \qquad (14.16)$$

$$= E_{pot} + E_{ZPE} + Nk_B T \left(T \frac{\partial \ln q}{\partial T} + 1\right)$$

---

1) Named for James Stirling (1692–1770), a Scottish mathematician, the Stirling formula is: $\ln x! = x \ln x - x$. Here, we use it as follows:

$$\ln Q = \ln\left(\frac{q^N}{N!}\right) = \ln q^N - \ln N! = N\ln q - (N\ln N - N)$$

Since N is a constant, the derivative is:

$$\left(\frac{\partial \ln Q}{\partial \beta}\right) = N\left(\frac{\partial \ln q}{\partial \beta}\right)$$

Using the chain rule, we obtain the result:

$$\left(\frac{\partial \ln q}{\partial T}\right) = \left(\frac{\partial \ln q}{\partial \beta}\right)\left(\frac{\partial \beta}{\partial T}\right) = \left(\frac{\partial \ln q}{\partial \beta}\right)\left(-\frac{1}{k_B T^2}\right)$$

and so

$$\left(\frac{\partial \ln q}{\partial \beta}\right) = -k_B T^2 \left(\frac{\partial \ln q}{\partial T}\right)$$

2) This is a critical point, since we calculate (macroscopic) thermodynamic properties using a single molecule only.

$$F(T) = U(T) - TS$$
$$= E_{\text{pot}} + E_{\text{ZPE}} - Nk_B T(\ln q + 1) \qquad (14.17)$$

and

$$G(T) = U(T) + pV - TS$$

$$= \underbrace{E_{\text{pot}} + E_{\text{ZPE}} + Nk_B T^2 \frac{\partial \ln q}{\partial T}}_{U}$$

$$+ \underbrace{Nk_B T}_{pV} - \underbrace{T Nk_B \left(T\frac{\partial \ln q}{\partial T} + \ln q + 1 \right)}_{S}$$

$$= E_{\text{pot}} + E_{\text{ZPE}} - Nk_B T \left(T\frac{\partial \ln q}{\partial T} + 1 - T\frac{\partial \ln q}{\partial T} - \ln q - 1 \right)$$

$$= E_{\text{pot}} + E_{\text{ZPE}} - Nk_B T \ln q \qquad (14.18)$$

heat capacity Finally, the heat capacity C_V is

$$C_V = \left(\frac{\partial U}{\partial T} \right)_{N,V} = Nk_B \left(2T\frac{\partial \ln q}{\partial T} + T^2 \frac{\partial^2 \ln q}{\partial T^2} \right) \qquad (14.19)$$

This is a very convenient result, as, at last, we only need to determine the molecular partition function q.

14.2.3
The Molecular Partition Function q

molecular The molecular partition function contains contributions from all
partition function degrees of freedom of a molecule resulting from translational, rotational, vibrational, and electronic degrees of freedom:

$$q = q^{\text{trans}} \times q^{\text{rot}} \times q^{\text{vib}} \times q^{\text{el}} \times \text{e},$$
$$\ln q = \ln q^{\text{trans}} + \ln q^{\text{rot}} + \ln q^{\text{vib}} + \ln q^{\text{el}} + 1 \qquad (14.20)$$

Euler's number (e) and the summand 1 ($\ln \text{e} = 1$) result from applying the Stirling formula in Equations 14.14 and 14.15 or at similar places for quantities other than the entropy. Both will be dealt with when calculating the translational partition function. As a typical example, we will follow the transformation for the entropy in the following. Try to do the same for the other quantities.

The entropy in Equation 14.15 can be transformed using Equation 14.20 in the following way:

$$S = Nk_B \left(T\frac{\partial \ln q}{\partial T} + \ln q + 1 \right)$$

$$= Nk_B \left[T\frac{\partial \ln(q^{\text{trans}} \times q^{\text{rot}} \times q^{\text{vib}} \times q^{\text{el}})}{\partial T} + \ln(q^{\text{trans}} \times q^{\text{rot}} \times q^{\text{vib}} \times q^{\text{el}}) + 1 \right]$$

$$= Nk_B \left[T\frac{\partial (\ln q^{\text{trans}} + \ln q^{\text{rot}} + \ln q^{\text{vib}} + \ln q^{\text{el}})}{\partial T} \right.$$

$$+ (\ln q^{\text{trans}} + \ln q^{\text{rot}} + \ln q^{\text{vib}} + \ln q^{\text{el}}) + 1 \Bigg]$$

$$= N k_{\text{B}} \Bigg[T \left(\frac{\partial \ln q^{\text{trans}}}{\partial T} + \frac{\partial \ln q^{\text{rot}}}{\partial T} + \frac{\partial \ln q^{\text{vib}}}{\partial T} + \frac{\partial \ln q^{\text{el}}}{\partial T} \right)$$

$$+ \ln q^{\text{trans}} + \ln q^{\text{rot}} + \ln q^{\text{vib}} + \ln q^{\text{el}} + 1 \Bigg]$$

$$= \underbrace{N k_{\text{B}} \left(T \frac{\partial \ln q^{\text{trans}}}{\partial T} + \ln q^{\text{trans}} + 1 \right)}_{S^{\text{trans}}} + \underbrace{N k_{\text{B}} \left(T \frac{\partial \ln q^{\text{rot}}}{\partial T} + \ln q^{\text{rot}} \right)}_{S^{\text{rot}}}$$

$$+ \underbrace{N k_{\text{B}} \left(T \frac{\partial \ln q^{\text{vib}}}{\partial T} + \ln q^{\text{vib}} \right)}_{S^{\text{vib}}} + \underbrace{N k_{\text{B}} \left(T \frac{\partial \ln q^{\text{el}}}{\partial T} + \ln q^{\text{el}} \right)}_{S^{\text{el}}}$$

$$= S^{\text{trans}} + S^{\text{rot}} + S^{\text{vib}} + S^{\text{el}} \qquad (14.21)$$

Therefore, all the quantities containing the molecular partition function can be written as sums

$$S = S^{\text{trans}} + S^{\text{rot}} + S^{\text{vib}} + S^{\text{el}}$$
$$U = E_{\text{pot}} + E_{\text{ZPE}} + E^{\text{trans}} + E^{\text{rot}} + E^{\text{vib}} + E^{\text{el}}$$
$$C_V = C_V^{\text{trans}} + C_V^{\text{rot}} + C_V^{\text{vib}} + C_V^{\text{el}} \qquad (14.22)$$

Note that the summand 1 is included in the translational part. Moreover, Equation 14.21 gives us the definitions of the different entropic parts (translational, rotational, vibrational, electronic). These will be used below. The single contributions are then calculated as follows.

Translation

Besides the molecular mass m, the translational contributions are determined by the temperature T and the pressure p. The volume V is calculated using the ideal-gas expression.[3] Thus, Equation 14.23 is only valid within the ideal-gas approximation:

translational contribution

$$q^{\text{trans}} = \left(\frac{2\pi m k_{\text{B}} T}{h^2} \right)^{3/2} V = \left(\frac{2\pi m k_{\text{B}} T}{h^2} \right)^{3/2} \frac{N k_{\text{B}} T}{p} \qquad (14.23)$$

---

3) From $pV = nRT$ it follows that

$$V = \frac{nRT}{p} = \frac{(N/N_{\text{A}})RT}{p} = \frac{N k_{\text{B}} T}{p},$$

with n, N, and N_{A} being the amount of substance, the number of particles (molecules), and Avogadro's number, respectively.

Using the derivative[4)]

$$\left(\frac{\partial \ln q^{\text{trans}}}{\partial T}\right)_V = \frac{3}{2T}$$

we obtain

$$S^{\text{trans}} = Nk_{\text{B}}\left(\ln q^{\text{trans}} + \frac{3}{2} + 1\right) \tag{14.24}$$

for the translational entropy. Note that this expression is only valid for indistinguishable molecules behaving as an ideal gas. For the translational part of the internal energy (cf. Equation 14.13) we find

$$E^{\text{trans}} = Nk_B T^2 \frac{\partial \ln q^{\text{trans}}}{\partial T} = Nk_B T^2 \left(\frac{3}{2T}\right) = \frac{3}{2}RT \tag{14.25}$$

Finally, the translational contribution to the heat capacity is (cf. Equation 14.25)

$$C_V^{\text{trans}} = \left(\frac{\partial E^{\text{trans}}}{\partial T}\right)_V = \frac{3}{2}R \tag{14.26}$$

Rotation

rotational contribution For the rotational contributions, we have to consider linear and non-linear molecules separately. For linear molecules, the rotational partition function is

$$q^{\text{rot}} = \frac{1}{\sigma}\frac{k_{\text{B}}T}{hcB} \tag{14.27}$$

and for non-linear molecules it is

$$q^{\text{rot}} = \frac{1}{\sigma}\left(\frac{k_{\text{B}}T}{hc}\right)^{3/2}\sqrt{\frac{\pi}{B_1 B_2 B_3}} \tag{14.28}$$

Here, B, B_1, B_2, and B_3 are the rotational constants. The symmetry number σ accounts for the number of indistinguishable orientations of the molecule. It is directly accessible from the point group (cf. Table 14.1).

The rotational constants B can be determined by the trace of the tensor of the moments of inertia. For simplification, the rotational temperature Θ^{rot} is introduced, which is also directly related to the trace

---

4) From Equation 14.23 we get $\ln q^{\text{trans}} = \frac{3}{2}[\ln(2\pi mk\,BV^{3/2}) + \ln T - \ln h^2]$ and so

$$\left(\frac{\partial \ln q^{\text{trans}}}{\partial T}\right)_V = \frac{3}{2T}$$

Table 14.1 Symmetry numbers of different point groups.

Point group			σ	Point group			σ	Point group	σ
C_1	C_i	C_s	1					$C_{\infty v}$	1
C_2	C_{2v}	C_{2h}	2	D_2	D_{2d}	D_{2h}	4	$D_{\infty h}$	2
C_3	C_{3v}	C_{3h}	3	D_3	D_{3d}	D_{3h}	6	T, T_h, T_d	12
C_4	C_{4v}	C_{4h}	4	D_4	D_{4d}	D_{4h}	8	O, O_h	24
C_5	C_{5v}	C_{5h}	5	D_5	D_{5d}	D_{5h}	10	I, I_h	60
C_6	C_{6v}	C_{6h}	6	D_6	D_{6d}	D_{6h}	12	S_4	2
C_7	C_{7v}	C_{7h}	7	D_7	D_{7d}	D_{7h}	14	S_6	3
C_8	C_{8v}	C_{8h}	8	D_8	D_{8d}	D_{8h}	16	S_8	4

of the tensor of the moments of inertia (I_i, $i = 1, 2, 3$ for the three spatial directions):

$$\Theta_i^{\text{rot}} = \frac{hc}{k_B} B = \frac{hc}{k_B} \frac{h}{8\pi^2 c I_i} = \frac{\hbar^2}{2 I_i k_B} \tag{14.29}$$

Equations 14.27 and 14.28 can now be transformed to

$$q^{\text{rot}} = \frac{1}{\sigma} \frac{T}{\Theta^{\text{rot}}} \tag{14.30}$$

and

$$q^{\text{rot}} = \frac{1}{\sigma} \sqrt{\frac{T^3}{\Theta_1^{\text{rot}} \Theta_2^{\text{rot}} \Theta_3^{\text{rot}}} \pi} \tag{14.31}$$

In analogy to the considerations for the translation we can again calculate the rotational contributions to the entropy, the internal energy, and the heat capacity using the rotational partition function. For linear molecules, these are

$$S^{\text{rot}} = N k_B (\ln q^{\text{rot}} + 1)$$
$$E^{\text{rot}} = N k_B T = RT$$
$$C_V^{\text{rot}} = R \tag{14.32}$$

and for nonlinear molecules, they are

$$S^{\text{rot}} = N k_B (\ln q^{\text{rot}} + \tfrac{3}{2})$$
$$E^{\text{rot}} = \tfrac{3}{2} N k_B T = \tfrac{3}{2} RT$$
$$C_V^{\text{rot}} = \tfrac{3}{2} R \tag{14.33}$$

Vibrations

Depending on its eigenfrequency or wavenumber, each single molecular vibration contributes to the partition function:

$$q^{\text{vib}} = \prod_{i=1}^{N_{\text{vib}}} q^{\text{vib, molecular}}(\bar{\nu}_i) \tag{14.34}$$

vibrational contribution The single molecular vibrational contributions $q^{\text{vib,molecular}}$ can be calculated individually from the wavenumber $\bar{\nu}_i$ of the ith vibration:

$$q^{\text{vib, molecular}}(\bar{\nu}_i) = \frac{1}{1-e^{-\beta hc\bar{\nu}_i}} = \frac{1}{1-e^{-\Theta_i^{\text{vib}}/T}} \tag{14.35}$$

using $\Theta_i^{\text{vib}} = hc\,\bar{\nu}_i/k_B$ and $\beta = 1/(k_B T)$, as above. The resulting contributions[5] to the entropy, internal energy, and heat capacity are

$$S^{\text{vib}} = Nk_B\left[T\left(\frac{\partial \ln q^{\text{vib}}}{\partial T}\right)_V + \ln q^{\text{vib}}\right]$$

$$= Nk_B\sum_{i=1}^{N_{\text{vib}}}\left\{\frac{\Theta_i^{vib}}{T(e^{+\Theta_i^{\text{vib}}/T}-1)} - \ln(1-e^{+\Theta_i^{\text{vib}}/T})\right\}$$

$$E^{\text{vib}} = Nk_B\sum_{i=1}^{N_{\text{vib}}}\frac{\Theta_i^{vib}}{e^{+\Theta_i^{\text{vib}}/T}-1} = R\sum_{i=1}^{N_{\text{vib}}}\frac{\Theta_i^{vib}}{e^{+\Theta_i^{\text{vib}}/T}-1}$$

$$C_V^{\text{vib}} = R\sum_{i=1}^{N_{\text{vib}}}e^{+\Theta_i^{\text{vib}}/T}\left(\frac{\Theta_i^{vib}}{T(e^{+\Theta_i^{\text{vib}}/T}-1)}\right)^2 \tag{14.36}$$

Electronic Contributions

electronic contribution Here, we will assume that all the molecules are in their electronic ground states. Electronic excitations do usually not occur thermally, since already a very small excitation of 0.1 eV would need a temperature of more than 1000 K. In this temperature range all other approximations will be problematic as well. Thus we will neglect the electronic contributions from the electronic degrees of freedom, that is,

$$q^{\text{el}} = 1 \quad \text{and} \quad \ln q^{\text{el}} = 0 \tag{14.37}$$

Therefore, the contributions to the entropy, internal energy, and heat capacity vanish.

14.2.4
Calculating the Relative Abundance of Isomers in the Gas Phase

If the activation barrier of an isomerisation reaction is sufficiently small with respect to $k_B T$, we can calculate the relative abundance of the

---

[5] Here, we have

$$\left(\frac{\partial \ln q^{\text{vib}}}{\partial T}\right)_V = \sum_{i=1}^{N_{\text{vib}}}\frac{\Theta_i^{vib}}{T^2(e^{+\Theta_i^{\text{vib}}/T}-1)}$$

Note that the product of contributions to the rotational partition function converts into a sum when the logarithmic laws are applied. Therefore, the derivative is a sum of derivatives as well, which makes the derivation easy. Note also that the exponential on the right-hand side is positive.

isomers in thermodynamic gas mixtures. In a system containing M **isomer mixture** compounds, the mole fraction x_i of the isomer i is

$$x_i(T) = \frac{q_i(T)e^{-\Delta U_i(0)/RT}}{\sum_{j=1}^{M} q_j(T)e^{-\Delta U_j(0)/RT}} \qquad (14.38)$$

14.3
Demonstration

Calculating Thermodynamic Properties of Methane

As an example, we will calculate the thermodynamic properties of methane at room temperature (298.15 K). Therefore, we first perform a geometry optimization of the CH_4 molecule, followed by a vibrational frequency analysis. If you use deMon as delivered with this book, you first use the OPTIMISATION keyword to perform a full geometry optimization. Your input may look like this:

```
TITLE CH4 OPT
VXCTYPE AUXIS PBE
BASIS (DZVP-GGA)
AUXIS (GEN-A2)
OPTIMISATION
GEOMETRY CARTESIAN ANGSTROM
C 0.0 0.0 0.0
H 0.7 0.7 0.7
H -0.7 -0.7 0.7
H 0.7 -0.7 -0.7
H -0.7 0.7 -0.7
END
```

The optimization yields values for the C—H bond lengths of 1.109 Å and an angle of 109.47°. After optimization, you change the guessed bond lengths and the angle in your input file to the optimized values, for example, in either Cartesian or internal coordinates. After that, you perform a vibrational analysis followed by the computation of some thermodynamic properties by removing the keyword OPTIMISATION and adding instead the two keywords FREQUENCY and THERMO.

Read through the output of this calculation. The vibrations are analyzed, the zero-point energy is calculated (26.8 kcal mol^{-1} = 0.0427 Hartree), and the symmetry number is given as well as the moments of inertia and the rotational constants (5.09798, 5.09791, and 5.09777 cm^{-1}). For the potential energy we will use the total energy (−40.4726 Hartree). The frequencies of the nine vibrations are stated as well. Some of these are degenerate (similar frequencies, similar intensities): 1288 cm^{-1} (three-fold degenerate), 1489 cm^{-1} (two-fold degenerate), 2895.2 cm^{-1}, and 3003 cm^{-1} (three-fold degenerate).

First, we calculate the four partition functions q^{trans}, q^{rot}, q^{vib}, and q^{el} using Equations 14.23, 14.28, 14.34, 14.35 and 14.37. Since we neglect any electronic contributions, we use $q^{\text{el}} = 1$. Besides the fundamental constants k_B, h, and c, we need the mass of the methane molecule ($m = M/N_A$), and the temperature (room temperature, $T = 298.15$ K), N is set to 1 (we use one molecule only), and we will use standard pressure ($p = p^{\ominus} = 1 \text{ atm} = 101325$ Pa). For q^{trans} and q^{rot} we can immediately use Equations 14.23 and 14.28:

$$q^{\text{trans}} = \left(\frac{2\pi \times 2.664 \times 10^{-26} \text{ kg} \times k_B \times 298.15 \text{ K}}{h^2} \right)^{3/2}$$

$$\frac{1 \times k_B \times 298.15 \text{ K}}{101325 \text{ Pa}} = 2.5256 \times 10^6$$

$$q^{\text{rot}} = \frac{1}{2} \left(\frac{k_B \times 298.15 \text{ K}}{hc} \right)^{3/2}$$

$$\sqrt{\frac{\pi}{509.798 \text{ m}^{-1} \times 509.791 \text{ m}^{-1} \times 509.777 \text{ m}^{-1}}} = 459.36$$

For q^{vib} we have to calculate the molecular contributions first using Equation 14.35. The vibrational frequencies of the methane molecule and the resulting rotational temperatures are given in Table 14.2. Note the degeneracy of the frequencies. We can also see that all vibrational contributions to the molecular partition function are close to unity, and so is their product (you can use a spreadsheet to recalculate them and to investigate the influence of the temperature on the partition function):

$$q^{\text{vib}} = \prod_{i=1}^{9} q^{\text{vib, molecular}}(\bar{\nu}_i) = 1.0075$$

We now calculate the entropy contributions according to Equations 14.24, 14.33, 14.36, and sum them up (Equation 14.22). The vibrational contributions are given in Table 14.2. Note that $S^{\text{el}} = 0$. We obtain:

$$S^{\text{trans}} = N k_B \left(\ln q^{\text{trans}} + \tfrac{3}{2} + 1 \right) = 143.36 \text{ J mol}{-1} \text{ K}^{-1}$$

$$S^{\text{rot}} = N k_B \left(\ln q^{\text{rot}} + \tfrac{3}{2} \right) = 63.44 \text{ J mol}^{-1} \text{ K}^{-1}$$

Table 14.2 Vibrational frequencies of the methane molecule and calculated vibrational properties. For the entropy, the heat capacity, and the internal energy contributions, the factor R is not included here.

i	$\bar{\nu}_1$ (m^{-1})	$\Theta_i^{vib}(\bar{\nu}_i)$ (K)	$q_{vib}(\bar{\nu}_i)$	S_{vib} (J mol^{-1} K^{-1})	C_V (J mol^{-1} K^{-1})	E_{vib} (J mol^{-1})
1	1287.8	1852.86	1.00200	1.44572×10^{-2}	7.75573×10^{-2}	3.713
2	1288.4	1853.72	1.00200	1.44211×10^{-2}	7.74043×10^{-2}	3.704
3	1289.8	1855.74	1.00198	1.43373×10^{-2}	7.70482×10^{-2}	3.683
4	1488.9	2142.20	1.00076	6.20777×10^{-3}	3.91848×10^{-2}	1.625
5	1489.3	2142.77	1.00076	6.19725×10^{-3}	3.91301×10^{-2}	1.622
6	2895.2	4165.55	1.00000	1.28112×10^{-5}	1.67034×10^{-4}	0.004
7	3002.9	4320.51	1.00000	7.88309×10^{-6}	1.06860×10^{-4}	0.002
8	3003.0	4320.65	1.00000	7.87953×10^{-6}	1.06816×10^{-4}	0.002
9	3003.1	4320.80	1.00000	7.87597×10^{-6}	1.06771×10^{-4}	0.002

$$S^{vib} = Nk_B \sum_{i=1}^{N_{vib}} \left\{ \frac{\Theta_i^{vib}}{T(e^{+\Theta_i^{vib}/T} - 1)} \right.$$

$$\left. -\ln\left[1 - e^{-\Theta_i^{vib}/T}\right] \right\} = 0.4628 \text{ J mol}^{-1} \text{ K}^{-1}$$

$$S = S^{trans} + S^{rot} + S^{vib} + S^{el} = 207.26 \text{ J mol}^{-1} \text{ K}^{-1}$$

The heat capacity C_V is calculated according to Equations 14.22 and 14.36. Note that the electronic contribution is neglected ($C_V^{el} = 0$) and the translational and rotational contributions both are $\frac{3}{2}R$ (Equations 14.26 and 14.34), so that we only have to calculate the vibrational contributions (see Table 14.2):

$$C_V^{vib} = R \sum_{i=1}^{3} e^{+\Theta_i^{vib}/T} \left(\frac{\Theta_i^{vib}}{T[e^{+\Theta_i^{vib}/T} - 1]} \right)^2$$

$$= 2.58 \text{ J mol}^{-1} \text{K}^{-1}$$

and

$$C_V = \frac{3}{2}R + \frac{3}{2}R + 2.58 \text{ J mol}^{-1} \text{ K}^{-1} = 27.53 \text{ J mol}^{-1} \text{ K}^{-1}$$

To calculate the internal energy we use the zero-point energy and the total energy obtained from the density-functional theory calculation: $E_{pot} = -40.4726$ Hartree and $E_{ZPE} = 0.0427$ Hartree. Moreover, $E^{trans} = E^{rot} = \frac{3}{2}RT = 0.0014$ Hartree. Since the electronic contribution is neglected, we just end up with calculating the vibrational contribution (see Table 14.2) and summing up (note that the different

Table 14.3 Results of the demonstration converted into different energy units.

	Hartree	eV	kJ mol^{-1}	J
Internal energy U	−40.4697	−1101.22	−106240.0	−1.7642×10^{-16}
Enthalpy H	−39.5254	−1075.53	−103761.0	−1.7230×10^{-16}
Free energy F	−40.4933	−1101.86	−106301.8	−1.7652×10^{-16}
Free enthalpy G	−41.4376	−1127.56	−108780.8	−1.8064×10^{-16}

contributions differ in orders of magnitudes):

$$E^{\text{vib}} = R\sum_{i=1}^{N_{\text{vib}}} \frac{\Theta_i^{\text{vib}}}{e^{+\Theta_i^{\text{vib}}/T}-1} = 4.5 \times 10^{-5} \text{ Hartree}$$

$$U = E_{\text{pot}} + E_{\text{ZPE}} + E^{\text{trans}} + E^{\text{rot}} + E^{\text{vib}} + E^{\text{el}} = -40.4697 \text{ Hartree}$$

The enthalpy at $T = 298.15$ K can now easily be calculated as
$H(T) = U(T) + RT = -39.5254$ Hartree,
the free energy is $F(T) = U(T) - TS(T) = -40.4933$ Hartree,
and the free enthalpy is $G(T) = F(T) - RT = -41.4376$ Hartree.
For the energy calculations, we have used Hartree as the energy unit. Table 14.3 converts all the energetic results from Hartree into eV, kJ mol^{-1} and J.

14.4
Problems

1. Thermodynamic properties of carbon dioxide

Calculate the internal energy, the enthalpy, the entropy, the free energy, the free enthalpy, and the specific heat capacity of carbon dioxide at 25 °C and 100 °C.

2. Reaction enthalpy

a) Calculate the reaction enthalpy and the reaction free enthalpy for the following reaction between ethene and hydrogen at 200 °C:

$$H_2C = CH_2 + H_2 \rightarrow H_3C - CH_3$$

b) In which temperature range do the reaction partners react spontaneously?

c) Compare your results to experimental data.

3. Relative abundances of difluoroethene isomers

Calculate the relative thermodynamic abundances of an isomeric gas mixture of *cis*- and *trans*-difluoroethene in the temperature range between 50 K and 5000 K. Plot your results as a function of temperature.

Hints

1. Note that carbon dioxide is a linear molecule.
2. Use the THERMO scripts as described in the Appendix.
3. For plotting functions (for example, the results with respect to temperature), you may use Xmgrace or GNUplot (see Appendix).

14.5
Review and Summary

In this chapter, we have connected quantum chemical calculations and statistical thermodynamics. The main approximation was to use a single molecule in the gas phase, that is, in vacuum, in order to get access to the molecular partition function. From this and the contributions from the different degrees of freedom, we were then able to calculate thermodynamic properties, such as internal energy, enthalpy, free energy, free enthalpy, entropy, and heat capacity. All these have been calculated within the ideal-gas approximation and using the harmonic approximation to treat the molecular vibrations.

In the Demonstration we have seen that the energetic contributions come mainly from the potential energy, and that, with respect to this, the other contributions are rather small. This is the reason why, for sufficiently low temperatures, the enthalpy equals the internal energy.

Terms like partition function, rotational constants, rotational temperature, and symmetry number have been introduced and discussed briefly. We have also seen that there are orders of magnitude between the different contributions to the molecular partition function, with the translational contribution being the largest and those from vibrations being the smallest. The electronic contributions have been neglected. Finally, you have calculated the relative abundance of the isomers in an isomeric gas mixture as a function of temperature.

References

1 See the chapter on "Statistical thermodynamics" in Atkins, P.W. and de Paula, J. (2006) *Physical Chemistry,* 4th edn, Oxford University Press.

2 See the white paper on the Gaussian web site: Ochterski, J. Thermochemistry in Gaussian (www.gaussian.com).

15
Molecular Dynamics – Basic Concepts

15.1
Aim

This experiment makes you familiar with the basic concepts of molecular dynamics, a powerful method for the computer simulation of molecular processes on the atomic scale. We study algorithms for the propagation of the molecular coordinates in time and understand the physical quantities of conservation. The importance of an adequate numerical treatment is discussed. A connection to the vibrations of small molecules is made. The concepts are introduced for a diatomic molecule, described with the Morse potential, where the simulation can be carried out using a spreadsheet.

15.2
Theoretical Background

15.2.1
Computer Simulations in Chemistry

Computer simulations allow the understanding of processes on time and length scales that are often beyond today's experimental accessibility. In the simulations, fundamental laws are applied to model systems. The results can be compared to experiment with the aim of understanding elementary processes in a better way, the basis for the improvement of systems for real applications.

computer simulations

Today, we are familiar with various types of computer simulations, each of them useful for a certain type of application. We will restrict ourselves here to dynamic simulations – just one of the simulation types that are useful in computational chemistry. In dynamic computer simulations, a system, given with certain starting conditions, is propagated in time. Applied to chemistry at the atomic scale, this means that a certain molecular structure is given, velocities are given or randomly

Computational Chemistry Workbook: Learning Through Examples
Thomas Heine, Jan-Ole Joswig, and Achim Gelessus
Copyright © 2009 WILEY-VCH Verlag GmbH & Co. KGaA, Weinheim
ISBN: 978-3-527-32442-2

assigned at the start, and the positions of the individual atoms are then propagated in time. We can observe the change of structure in time, and we may spot certain events, such as rearrangement of atoms or even chemical reactions. Such simulations are commonly called **molecular dynamics** molecular dynamics (MD) simulations. In brief, for classical MD simulations, the system is given at time t by positions $\{R_I\}$ and velocities $\{\partial R_I / \partial t = \dot{R}_I\}$ of all atoms $I = 1, 2, \ldots, N$, forces on all atoms are calculated as negative derivatives of the potential energy surface with respect to the coordinates $F_I = -\Delta_I V$, and using this information and Newton's second law

$$F_I = m_I \frac{\partial^2 R_I}{\partial t^2} = m_I \ddot{R}_I$$

a new set of positions and velocities at time $t + \Delta t$ is calculated, and the simulation is continued for as many steps as necessary – or possible.

quantum MD MD simulations are usually very intensive in computer time. Therefore, different levels of approximations have been established, which cover different levels of accuracy and can be applied to study systems at different length and time scales. We can classify them by distinguishing the classical from the quantum mechanical treatment. If the nuclei and the electrons are treated quantum mechanically, we speak of quantum molecular dynamics (QMD). The quantum mechanical description of the motion of the nuclei is in most cases not necessary, as quantum effects are usually small, except, as we have learned in Chapter 4, in some situations for protons. An exception is accurate thermochemistry (see the discussion of ZPE in Chapters 4 and 14), but this subject is not the main goal of MD simulations. Since QMD simulations require an enormous amount of computer time, the time scales and system sizes that can be treated even with modern computers are rather small. It is therefore usually a good approximation to treat the nuclei classically, whereas the electronic system is still treated quantum mechanically. **Born–Oppenheimer MD** This approximation is called Born–Oppenheimer molecular dynamics (BOMD), because we are applying the Born–Oppenheimer approxi-**Car–Parrinello MD** mation. A very efficient approximation to BOMD is the so-called Car–Parrinello molecular dynamics (CPMD) [1].[1] Finally, the electronic system may also be treated classically, for example, in terms of a classical force field that contains all the information we know about the chemical behavior of the elements, like typical bond lengths, bond angles, and dihedral angles. In this case we speak of classical molecular dynamics (CMD). CMD simulations of this kind may target interesting subjects such as protein folding or diffusion rates, but usually not

---

1) In Car–Parrinello molecular dynamics, the molecular orbitals are treated in the same way as nuclei and propagated in time. This simulation technique was a breakthrough in 1985 and opened the field of predictive computer simulations in chemistry and condensed matter physics.

processes where chemical reactions are involved. In this experiment we will concentrate on CMD and BOMD.

If we were to be exact in our nomenclature, BOMD (and CPMD) fall into the field of CMD, as the equations of motion of the nuclei are treated using classical mechanics. However, the literature is quite ambiguous in the usage of these terms. The term CMD, or even only MD, is commonly used if the electrons are not treated explicitly, but classical force fields are used. We follow this nomenclature here. On the other hand, QMD is frequently used for BOMD or CPMD, as the electronic system is treated using a quantum mechanical Hamiltonian. We will use the unambiguous term BOMD in this context. The bottom line is that we should always define what we mean when using the terms QMD and CMD.

15.2.2
The Born–Oppenheimer Approximation

The Born–Oppenheimer approximation, already mentioned in Chapter 9, separates the degrees of freedom of a molecule: electrons and nuclei are treated independently.[2] For the electronic system, it is assumed that the nuclei are frozen. This is usually a very good model, because the mass of an electron is about 2000 times lower than that of a proton, which means that the velocity of the nuclei is significantly smaller than that of the electrons.

Born–Oppenheimer approximation

Therefore, at each time step, the stationary Schrödinger equation of the electronic system is solved in the external potential of the nuclei, frozen in their positions at the corresponding time step. The result is the total energy of the electronic system and its gradient. This is the essence of BOMD.

15.2.3
The Trajectory: Startup Conditions and Propagation of Atoms in Time

The trajectory is the "recording" of all positions of the particles during the simulation and is given by the continuous set of functions $\{R_I(t)\}$, defined between the starting point and the end of the simulation. The trajectory is defined solely by the starting conditions $\{R_I(t_0); \dot{R}_I(t_0)\}$ and by the approximation to treat the system, which means the type (and quality) of the force field for a CMD, or of the type of Hamiltonian for

---

2) Note that the synonymous usage of nuclei and atoms is incorrect. In the Born-Oppenheimer approximation, we separate nuclei and electrons to solve the electronic Schrödinger equation. The resulting forces have to be applied to the whole system, which includes nuclei and electrons. Strictly speaking, this is wrong, as we do not know how many electrons need to be associated to an atom. The large mass difference between electrons and nuclei, however, justifies to use the atomic mass.

BOMD. A further, important factor is the way in which the trajectory is calculated numerically. After the trajectory of the system is calculated, a lot more is known about the behavior of the system, and, by applying statistics, various properties of the system can be calculated. For details we refer the reader to the literature [2, 3].

15.2.4
The Verlet Algorithm

For the computation of a trajectory, we need to calculate the future. Assume that we are at some time t in the trajectory. From this point, defined as $\{R_I(t); \dot{R}_I(t)\}$, the positions and velocities in the future, at $t + \Delta t$, $\Delta t > 0$, need to be calculated. To do so, we may expand the continuous functions $\{R_I(t)\}$ in t, with an advancement of Δt:

$$R_I(t+\Delta t) = R_I(t) + \dot{R}_I(t)\, \Delta t + \frac{1}{2}\, \ddot{R}_I(t)(\Delta t)^2 + \cdots \qquad (15.1)$$

Equation 15.1 terminates with the second-order term, as this is the last term that we can calculate directly using the laws of physics. Newton's second law connects the acceleration with the force acting on the particle, which is conveniently calculated by the derivative of the potential with respect to the atomic positions of atom I:

$$\ddot{R}_I(t) = \frac{\mathbf{F}_I(t)}{m_I} = -\frac{\nabla_I V(\tilde{t})}{m_I} = -\frac{1}{m_I}\begin{pmatrix} \partial/\partial X_I \\ \partial/\partial Y_I \\ \partial/\partial Z_I \end{pmatrix} V(R_1, R_2, \ldots, R_N, \tilde{t})$$

$$(15.2)$$

We can use Equation 15.1 to calculate the trajectory, but this is known to be an inefficient algorithm. Verlet approached the result from two sides, writing down the Taylor series of $\{R_I(t)\}$ to predict the future, which is Equation 15.1, but also that to calculate the past:

$$R_I(t-\Delta t) = R_I(t) - \dot{R}_I(t)\, \Delta t + \frac{1}{2}\ddot{R}_I(t)(\Delta t)^2 - \cdots \qquad (15.3)$$

Verlet algorithm   If one adds Equations 15.1 and 15.3, one obtains the Verlet algorithm:

$$R_I(t+\Delta t) = 2R_I(t) - R_I(t-\Delta t) + \ddot{R}_I(t)(\Delta t)^2 + \cdots \qquad (15.4)$$

Equation 15.4 is much more accurate and more convenient than Equation 15.1. First, the error is reduced significantly. Equation 15.1 has an intrinsic error that goes with $(\Delta t)^3$, as the third and higher terms of the Taylor expansion are neglected. In the addition of Equations 15.1 and 15.3 the third-order terms cancel each other exactly, so Equation 15.4 carries only an intrinsic error going with $(\Delta t)^4$. Second, the derivation makes the algorithm time reversible. This means that, if we can calculate the forces exactly, we can, at any point, reverse the velocities and return exactly to the starting position of our simulation.

This is indeed a nice test for the quality of a numerical implementation of the method.

For the calculation of the next step of the trajectory, the Verlet algorithm requires the positions of the present and of the previous step. Since it is a simple and robust mechanism, it is very often applied. In the forthcoming, we will calculate the trajectory in an iterative manner. This means that we will write the simulation time of iteration n, $n = 0,1,2,\ldots,n_{max}$, as t_n, and the next iteration is $t_n + \Delta t = t_{n+1}$. That is, we always use equidistant time steps. The Verlet algorithm transforms this way to

$$R_I^{(n+1)} = 2R_I^{(n)} - R_I^{(n-1)} + \ddot{R}_I(n)(\Delta t)^2 \tag{15.5}$$

We note the iteration as superscripts in parentheses to avoid confusion with the subscript, which denotes the index of the nucleus.

15.2.5
The Velocity Verlet Algorithm

The Verlet algorithm has, however, one drawback: one cannot calculate the velocities at the present step. Velocities are not part of the algorithm, and hence neither are they propagated, nor does any velocity information enter Equation 15.5. This is, at first glance, surprising, but obviously the information is contained in the fact that positions at two consecutive time steps are needed in the Verlet algorithm. The velocities can be calculated a posteriori, that is for any previous step, by

$$\dot{R}_I^{(n)} = \frac{R_I^{(n+1)} - R_I^{(n-1)}}{2\Delta t} \tag{15.6}$$

This drawback may be irrelevant for many simulations. It is, however, convenient to know the velocities at the simulation time, for example to calculate the kinetic energy – and hence the temperature – of the system.

A second inconvenience is the problem that we cannot start off the simulation from the starting conditions $\{R_I^{(0)}; \dot{R}_I^{(0)}\}$, but we have to transform them to two sets of coordinates $\{R_I^{(0)}; R_I^{(-1)}\}$.

A slight modification of the Verlet algorithm leads to the velocity Verlet algorithm. Here, the positions are propagated as known from the Taylor expansion, which is cut off after the second term:

velocity Verlet algorithm

$$R_I^{(n+1)} = R_I^{(n)} - \dot{R}_I^{(n)}\Delta t + \frac{1}{2}\ddot{R}_I^{(n)}(\Delta t)^2 \tag{15.7}$$

The velocities are then calculated as

$$\dot{R}_I^{(n+1)} = \dot{R}_I^{(n)} + \frac{1}{2}(\ddot{R}_I^{(n)} + \ddot{R}_I^{(n+1)})\Delta t \tag{15.8}$$

Note that the accelerations for step $n + 1$ can be calculated using Newton's second law after application of Equation 15.7. In practice, it is not convenient to store two sets of accelerations. Therefore, the velocity Verlet algorithm is usually applied as shown in the following procedure.

1. Calculate the new positions (this is Equation 15.7):

$$R_I^{(n+1)} = R_I^{(n)} - \dot{R}_I^{(n)} \Delta t + \frac{1}{2} \ddot{R}_I^{(n)} (\Delta t)^2$$

2. Calculate the velocities at the next half-step, that is, at $t + \frac{1}{2} \Delta t$:

$$\dot{R}_I^{(n+\frac{1}{2})} = \dot{R}_I^{(n)} + \ddot{R}_I^{(n)} \frac{1}{2} \Delta t \qquad (15.9)$$

3. Compute the new accelerations $\{\ddot{R}_I^{(n+1)}\}$ using the positions $\{\mathbf{R}_I^{(n+1)}\}$ calculated in step 1.

4. Compute the new velocities using the results of steps 2 and 3 as:

$$\dot{R}_I^{(n+1)} = \dot{R}_I^{(n+\frac{1}{2})} + \ddot{R}_I^{(n+1)} \frac{1}{2} \Delta t \qquad (15.10)$$

5. Start over from step 1 for the next iteration.

We will use the Verlet algorithm in the Demonstration and the velocity Verlet algorithm in one of the problems. The latter is also used in most of the modern computational chemistry software, including the deMon program as delivered with this book.

15.2.6
Conservation of Fundamental Physical Quantities: Energy, Momentum, and Angular Momentum

We know from general physics that various quantities are *conserved*; this means that, in a closed system, these quantities cannot change. In the classical mechanics of point mass systems, these quantities are the total energy, the total momentum and the total angular momentum.

total energy   The total energy is defined by potential and kinetic energy of the system:

$$E = \sum_{I=1}^{N} \frac{1}{2} m_I \dot{R}_I^2 + E_{\text{pot}} = \text{constant} \qquad (15.11)$$

where the potential energy is defined by the potential energy surface of the present position, and thus it contains the electronic energy and the nucleus–nucleus repulsion energy. The two quantities are not constant in time – if we think of the oscillation of a diatomic molecule, the kinetic energy would be zero at the turning points, while it will have a maximum at the location of lowest potential energy.

The total momentum is defined as

$$P = \sum_{I=1}^{N} m_I \dot{R}_I = \text{constant} \equiv 0 \qquad (15.12)$$

Usually, the total momentum is set to zero, as this avoids the situation in which the center of mass of the structure moves during the simulation.

Finally, the total angular momentum, which is defined as

$$L = \sum_{I=1}^{N} m_I R_I \times \dot{R}_I = \text{constant} \equiv 0 \qquad (15.13)$$

is set to zero on the same grounds: otherwise, the molecule is permanently rotating.

It is reasonable to start molecular dynamics simulations in such a way that total momentum and total angular momentum are zero.

15.2.7
Numerical Considerations

For molecular dynamics simulations, we have the advantage that we can control the quality of the trajectory by checking the constancy of those quantities which need to be conserved by reasons of physics, as discussed above. If these quantities are not constant, then we must check the following possibilities.

- Is the system really closed? That is, are all types of energy transfer avoided, for example due to a thermostat (see Chapter 16), an external field (electric or magnetic), or a continuation of the system, for example, to account for the solvent or the spatial extension of a crystal?
- Is the time step reasonable? A good estimate for the time step is 0.25 fs for a hydrogen atom in a simulation of 300 K. Heavier atoms may allow larger time steps – the temperature is directly connected to the kinetic energy, and the velocity estimate for the proton can be multiplied by the square root of the value of the mass of the molecule's heaviest atom. Similarly, the time step needs to be decreased for higher temperatures. Note that for very long simulations even smaller time steps are necessary, as errors accumulate during the simulation.
- Are there inaccuracies in the computational approach to calculate the gradients? This happens particularly often in the case of Born–Oppenheimer molecular dynamics, and it is generally a very common point in MD simulations. The accuracy of the forces on the atoms are essential for the numerical stability of the approach – they are much higher than, for example, in a geometry optimization.

15.2.8
Heat Bath: Thermostats

It is not straightforward to start a molecular dynamics simulation so that it will correspond to the attempted simulation conditions. It is particularly difficult to set the velocities: at the starting point, velocities might be set to zero (if the molecule is not in the equilibrium position), or a random velocity distribution can be assigned to the nuclei. It is not clear, however, what average temperature this velocity distribution will account for. It is therefore important to allow a heat transfer to the system to adjust the average temperature to the given value. The numerical tool suiting this purpose is called a thermostat. In this experiment we shall not use such a tool, and we refer to Chapter 16 for its explanation.

15.3
Demonstration

15.3.1
Classical Molecular Dynamics in a Spreadsheet

We will apply the Verlet algorithm in a spreadsheet for a diatomic molecule. We can make use of the spreadsheet which we have already developed for the Demonstration of Chapter 9. Again, we will apply the Morse potential for the description of the potential energy surface (see Equation 3.1):

$$V(R) = D_e[(1 - e^{-\alpha(R-R_e)})^2 - 1] \tag{15.14}$$

Besides the potential, you may also use the expression of the force that you have already programmed for Chapter 9:

$$F(R) = -\frac{\partial V(R)}{\partial R} = -2D_e\alpha(e^{-\alpha(R-R_e)} - e^{-2\alpha(R-R_e)}) \tag{15.15}$$

As before, we use the carbon monoxide molecule as a test case. The necessary Morse parameters can be found in Table 3.1: $D_e = 1072.8\ \text{kJ mol}^{-1}$, $\alpha = 2.312 \times 10^{-2}\ \text{pm}^{-1}$, and $R_e = 113\ \text{pm}$. The reciprocal mass of CO is

$$\mu = \left(\frac{1}{M_C} + \frac{1}{M_O}\right)^{-1} = 6.86\ \text{g mol}^{-1}$$

We want to simulate the trajectory of this molecule if we start the simulation with a frozen molecule at an intermolecular distance of our choice, for example, for $R^{(0)} = 100$ pm.

We will first have to program a spreadsheet for the Verlet algorithm. The first column is the step number n. It is convenient to start with step -1, as the Verlet algorithm requires positions of the two last consecutive steps. The second column contains the time, which is $n\Delta t$. The third column could be the position. Here we place the starting value for steps -1 and 0. The fourth, fifth, ... columns could be the potential, then the force, then the acceleration, then the velocity, and finally the kinetic energy, and the total energy. It is convenient to use the units kJ mol^{-1} for energies, pm for distances, and fs for time. In this way all the units are consistent with each other and no further transformations have to be carried out.

We can now compute the values of the spreadsheet as follows. The potential $V^{(n)}$ is Equation 15.14, which requires a couple of constants and the value for the interatomic distance $R^{(n)}$ (in the same line). The force $F^{(n)}$ is calculated using Equation 15.15; again, it only requires $R^{(n)}$. For the acceleration, we divide $F^{(n)}$ by the reduced mass, in this case 6.86 g mol^{-1}. The velocity needs to be calculated from the positions of the previous and of the next time step, as given in Equation 15.6. We can therefore calculate it only for the zeroth (if we start with $n_{\text{start}} = -1$) until the second last time step of the simulation, as information on $n - 1$ and $n + 1$ is necessary. The kinetic energy is, for the transformed coordinates (see Chapter 9), given by $E_{\text{kin}}^{(n)} = \frac{1}{2}\mu(\dot{R}^{(n)})^2$, and the total energy is the sum of kinetic energy and potential.

We start with the Verlet algorithm for $n = 1$, for which we program Equation 15.5. Now we can use the copy function of the spreadsheet and copy the iterative procedure for many steps, for example, for $n_{\text{max}} = 1000$. If everything is done correctly, we should see the trajectory of the diatomic molecule in our spreadsheet – we have written a simple CMD code.

Let us start with an initial geometry of $R^{(-1)} = R^{(0)} = 100$ pm and a time step of $\Delta t = 0.25$ fs. For the first row we obtain the following values. The potential is $V^{(0)} = -940.9198$ kJ mol^{-1}. The acceleration is 3.4240 pm fs^{-2}. The total energy is, as the velocity is zero, identical to the potential energy. The position at the first step is $R^{(1)} = 100.21400$ pm, leading to a potential of 945.88677 kJ mol^{-1}. The first meaningful velocity is calculated at step 2, for which we obtain a value of 1.274 pm fs^{-1}.

If we move down the spreadsheet, we see that the values for the velocity and acceleration are changing signs, and that the potential and kinetic energies change their values significantly. As required by physics, the total energy appears to remain constant. We observe an average of -940.67135 kJ mol^{-1}. We can plot potential, kinetic energy,

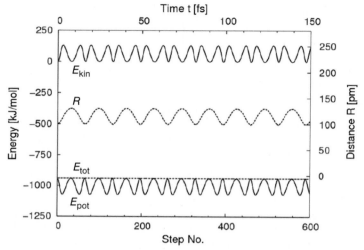

Figure 15.1 Results of an MD simulation for a CO molecule described by a Morse potential: kinetic (E_{kin}), potential (E_{pot}), and total (E_{tot}) energies (kJ mol^{-1}) and interatomic distance (pm) as functions of simulation step number and simulation time (in fs), respectively.

and total energy, as shown in Figure 15.1. We observe oscillations: obviously potential and kinetic energy oscillate with the same frequency and amplitude. However, they are phase-shifted – whenever the potential has a minimum, the kinetic energy has a maximum, and vice versa.

We may also discuss the form of the potential energy curve. We do not see a symmetric sine curve, as we would have expected from a harmonic potential. The form is given by the asymmetry of the Morse potential (see Figure 3.1). At interatomic distances smaller than the equilibrium distance, the steeper potential gives stronger reset forces, and the system's dwell time is shorter. The oscillation is carried on in time – we have learned everything about the system after the first full cycle. The reason for this "boring" behavior is that we have only one degree of freedom, and the molecule can do nothing other than oscillate for eternity, since the total energy is constant.

But is that really the case? Let us inspect the numbers for the total energy more closely. Even if we use only four significant digits – this is 0.1 kJ mol^{-1} – we see that the total energy is not constant, but oscillating. We can also plot the total energy separately on an appropriate energy scale (Figure 15.2) to see the oscillations. We know that a large part of this error comes from the estimation of the velocity, which enters the kinetic energy quadratically. Indeed, a reduction of the time step to $\Delta t = 0.1$ fs moves the oscillations of the total energy to the second

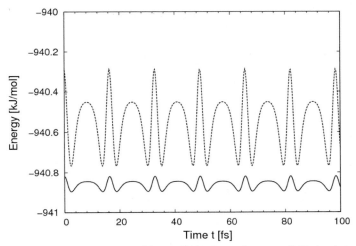

Figure 15.2 The total energy of the CO molecule as a function of MD simulation time for two different time steps: $\Delta t = 0.1$ fs (solid curve) and $\Delta t = 0.25$ fs (dashed curve). Note that the difference in average total energy is related to the starting point: the error results in an offset of both trajectories. Further note that the fluctuation of the total energy is highest for the steep part of the potential, when changes in the velocity are large. Note also the scaling of the y axis in comparison to that of Figure 15.1, where the same curves appear as a constant function.

decimal place. Whereas the amplitude of the oscillations with $\Delta t = 0.25$ fs is approximately $0.5 \, \text{kJ mol}^{-1}$, it is decreased to $0.1 \, \text{kJ mol}^{-1}$ for $\Delta t = 0.1$ fs. But is this small step size really necessary for the trajectory? If the total energy only oscillates due to an inaccurate estimate of the total energy, we still have a correct trajectory. We need to check, therefore, if the total energy is only oscillating around a fixed value, or if it has an obviously wrong behavior, such as a continuous drift to larger or smaller values. The constancy of the total energy, as seen in Figure 15.1, magnified in Figure 15.2, gives us confidence that the simulation is carried out with reasonable numerical details.

Let us now deliberately perform a calculation with unreasonable numerics. We put more energy into the system, by changing the starting distance to 90 pm, and choosing a time step of 2 fs. We observe the dissociation of the molecule, which is obviously an artifact, as the initial potential value is $-544.2 \, \text{kJ mol}^{-1}$, so the molecule is still well bound. If we reduce the time step to 1 fs, we observe oscillations. However, we have total energy changes in the $10 \, \text{kJ mol}^{-1}$ range, hardly convincing for a sound MD trajectory. If we reduce the time step further to 0.25 fs or (better) 0.1 fs, we obtain a clean trajectory, without a significant drift of the total energy.

15.4
Problems

1. Molecular dynamics in a spreadsheet: test the numerics

Program the velocity Verlet algorithm in a spreadsheet for the carbon monoxide (CO) molecule. Compare the trajectories of both algorithms for the same starting conditions and the same time step, starting from 90, 100, 110, and 120 pm, and using a time step of 0.1, 0.25, 0.5, and 1 fs, respectively. Explain why the simulation with 90 pm intermolecular distance as starting condition requires a smaller time step than the simulation starting with a distance of 110 pm?

2. Molecular dynamics for HCl

Repeat the calculations of the Demonstration using the hydrogen chloride (HCl) molecule. Morse parameters can be found in Table 3.1. Remember to adjust the reduced mass.

3. Energetics in a molecular dynamics simulation

Use the equilibrium distance of hydrogen chloride (HCl) as startup geometry. What can be observed for the kinetic, the potential, and the total energies? Why? Now, use the velocity Verlet algorithm and give a startup velocity. Determine the startup velocity that is necessary to dissociate the molecule.

Hints

Use the formulas you have developed for your spreadsheed for the demonstration and problems of Chapter 9.

15.5
Review and Summary

In this computer experiment, we learned the basics of molecular dynamics. First we worked out the differences between classical and quantum molecular dynamics procedures. Hereby, we have to remember that sometimes these terms refer to the treatment of the nuclei and sometimes to that of the electrons. For the introduction of molecular dynamics, we have chosen the Born–Oppenheimer approach.

Generally, we need to know the positions of the atoms at $t = 0$ and their initial velocities to perform a molecular dynamics simulation. From these we can evolve the system in time. For the time evolution we have used two algorithms, the Verlet algorithm and the velocity Verlet algorithm. Especially the latter is used in standard quantum chemistry program packages to perform MD simulations. The Verlet algorithm combines an expression of the atomic positions at $t - \Delta t$ and at $t + \Delta t$, which results in a time-reversible expression without an explicit formulation of the velocities. However, the Verlet algorithm has the drawbacks that the velocities at the present step are unknown, and that we need to know the atomic positions of the last step already at the first step. The velocity Verlet algorithm, on the other hand, uses the velocities, so that the past does not have to be known.

We have also discussed the conservation of energy, momentum, and angular momentum, and raised several considerations on numerical issues. These could be checked in the Demonstration, where we programmed a basic classical MD code using a Morse potential to describe the interatomic interactions of a diatomic molecule with both a Verlet algorithm and a velocity Verlet algorithm for the time propagation. Since a diatomic molecule has only one degree of freedom, we could test our algorithm very easily, by, for example, plotting the potential, the kinetic, and the total energies. The oscillations in the first two arose from the oscillating behavior of our system, which was not in equilibrium at $t = 0$. We could observe the anharmonicity in the energy oscillations due to the asymmetric Morse potential. Finally, we could use the spreadsheet algorithm to see where the dissociation limit for our system is located and how a change in time step influences the oscillations.

References

1 Car, R. and Parrinello, M. (1985) Unified approach for molecular dynamics and density-functional theory. *Physical Review Letters*, **55**, 2471–2474.

2 Allen, M.P. and Tildesley, D.J. (1989) *Computer Simulation of Liquids*, Oxford University Press.

3 Frenkel, D. and Smit, B.J. (2002) *Understanding Molecular Simulation*, Academic Press.

16
Molecular Dynamics and Basic Thermodynamics

16.1
Aim

We will introduce two simulation ensembles: the microcanonical ensemble (constant number of particles, constant volume, constant system energy: NVE) and the canonical ensemble (constant number of particles, constant volume, constant system temperature: NVT). For this purpose we need the connection of the kinetic energy to the temperature, and a thermostat that establishes the connection between the simulated species and an external heat bath. We will discuss the conserved quantities in both ensembles and stress the importance of mean values in computer simulations.

16.2
Theoretical Background

In Chapter 15 we have learned the basics of a molecular dynamics simulation. One important result was that the total energy of the simulated system was dependent on the initial starting conditions (in our case on the interatomic distance of the diatomic molecule), and that this quantity was conserved within controllable numerical limits. The anharmonic oscillator, given by the Morse potential that was used for the MD simulation in Chapter 15, was just converting kinetic energy into potential energy and vice versa – the system was not able to lose or gain energy.

    In this chapter, we will now discuss how energy can be transferred to or from the system, and in this way we will make the link to one of the most important quantities of computer simulations, the temperature. The energy transfer is done by a so-called heat bath or thermostat, two terms that are known from macroscopic chemistry. In the case of a molecular dynamics simulation, these terms are used for numerical

molecular dynamics

thermostat

Computational Chemistry Workbook: Learning Through Examples
Thomas Heine, Jan-Ole Joswig, and Achim Gelessus
Copyright © 2009 WILEY-VCH Verlag GmbH & Co. KGaA, Weinheim
ISBN: 978-3-527-32442-2

procedures that serve the same purpose: to adjust the average[1] temperature to a given value. Before we go into detail, we will first review some basic statistical thermodynamics.

16.2.1
Basic Thermodynamics: Ensembles

The state of an ensemble of particles (in the laboratory) is defined through the exchange of different thermodynamic quantities with its **thermodynamic ensemble** environment. Different thermodynamic ensembles can therefore be defined as follows. For a detailed discussion we refer the interested reader to the common textbooks.

- **Grand canonical ensemble**: Particles and energy can be exchanged with the environment. Therefore, this type of ensemble is inapplicable for molecular dynamics simulations.[2]

- **Canonical ensemble**: Energy can be exchanged with the environment (the heat bath); the number of particles (N), the volume of the simulated domain (V), and the temperature (T) stay constant. Therefore, this ensemble is also called the *NVT* ensemble. As we will see **NVT ensemble** below, a thermostat is used in an MD simulation, that is, the exchange of kinetic energy is simulated by modifying the atomic velocities.

- **Microcanonical ensemble**: No energy exchange is allowed, that is, the total energy (E) consisting of potential energy and kinetic energy is a conserved quantity. As a consequence, volume (V) and number of particles (N) are also constant. Therefore, the microcanonical **NVE ensemble** ensemble is also called the *NVE* ensemble. During an exothermic process in a microcanonical ensemble, the simulation temperature rises; during an endothermic process, it decreases.

The definitions of the canonical and microcanonical ensembles introduced above are used, even though we have no definition for the volume in the case of treating a single molecule. In a gas-phase simulation we apply the ideal-gas approximation, which is equivalent to an infinite volume, and the terms *NVE* and *NVT* are also commonly used for such a type of simulation.

For *NVE* simulations, the constancy of the total energy is a convenient control parameter. A control parameter allows us to determine if the numerical conditions of a computer simulation are sound or not – if the control parameter does not remain constant, either the numerical parameters could have been chosen in a too sloppy way (e.g., too large

---

1) The role of the average temperature is explained in detail below.
2) The state-of-the art simulation technique of the grand canonical ensemble is the Grand Canonical Monte Carlo (GCMC) technique. We will not discuss Monte Carlo simulations in this workbook.

time step, too many approximations in the numerical method to solve the electronic Schrödinger equation), there could be a coupling to an external system, or the simulation could be subject to inherent errors, either in its methodology or in its implementation (usually a software bug).

If, on the one hand, we discuss an NVE ensemble, that is, we have a closed system, the total energy is given as the sum of the potential and the kinetic energy. This sum must be constant. If, on the other hand, we discuss an NVT ensemble, the total energy is not constant, as energy can be transferred to or from the system, which is connected to an external heat bath. We know from general physics that the temperature of a system is directly related to the motion of its atoms. In molecular dynamics simulations, we have direct access to the velocities of the individual atoms, and they determine the kinetic energy of the system (see Chapter 15) and thus the system's temperature.

16.2.2
The Temperature and the Ergodic Theorem

For a *reasonably large system* we can relate the temperature T to the kinetic energy, which itself is related to the velocities of the individual atoms:

$$T = \frac{2E_{\text{kin}}}{f k_{\text{B}}} = \frac{\sum_{I=1}^{N} m_I \, \dot{R}_I^2}{f k_{\text{B}}} \tag{16.1}$$

where f denotes the number of degrees of freedom, which is either $3N - 6$ ($3N - 5$ for linear systems, N being the number of atoms), or $3N$ if we also choose to account for the energy in the external degrees of freedom of the particles. The square of the velocities is understood as the square of the magnitude of the velocity vector, $\dot{R}_I^2 = \dot{R}_I \cdot \dot{R}_I = |\dot{R}_I|^2$. The principal question is this: Which system size is "reasonably large."

In the MD simulation of the diatomic carbon monoxide molecule in Chapter 15, you would certainly have realized that the kinetic energy was not (by any means) constant during the simulation, and hence the temperature would not have remained constant. Instead, it would have oscillated in the same way as the kinetic energy if defined by Equation 16.1. Assume, however, that you had simulated many independent particles instead: an ideal gas of CO molecules. You would have started the simulations independently at different times (e.g., by getting the startup time by a random number generator), and once all of them have started to run you calculate the system's kinetic energy, which is the sum of all the kinetic energies of all the individual CO molecules. If the number of starting points is indeed sufficient and the setups have been chosen independently, you will observe that the

ensamble average

fluctuation of the kinetic energy of the system reduces with the number of simulations, and it gets closer and closer to a constant value if you use more and more molecules. If you inspect Equation 16.1 you will realize that the temperature is nothing other than a prefactor times the average of the square of the velocities, and for an infinite number of independent molecules the average must be a constant temperature for any snapshot $\{(\dot{R}_I(t))^2\}$.

The temperature is a quantity that is independent of the system size, and can be considered as an average kinetic energy per particle, multiplied by a constant. In nature, the macroscopic system (of which we can measure the temperature) is in the order of 10^{23} molecules and Equation 16.1 is valid to an excellent approximation. For a computer simulation, the relation is less obvious. A reasonable system size depends strongly on the character of the system, and for practical simulations we can be fairly sure that the system size is too small to interpret T of Equation 16.1 as temperature. How can we escape this dilemma?

time average

Obviously, all individual simulations of the N CO molecules contain the same information, but have different starting conditions. It must, therefore, be possible to obtain the same information from the trajectory of a single molecule. Indeed, we must arrive at the same result if we carry out one single-molecule molecular dynamics trajectory and calculate the temperature by a series of snapshots taken at different, independent time steps. The summation index in Equation 16.1 will then run over the time steps, and will not refer to the different systems. In this way, much longer summations (large upper limit N) are usually possible.

In statistical physics we use a very similar approach to the one given above, which becomes equivalent for large summations. We define the temperature as a quantity proportional to the mean kinetic energy per particle, which is averaged over a certain simulation time t:

$$T = \frac{2\langle E_{\text{kin}} \rangle_t}{f k_{\text{B}}} = \frac{\left\langle \sum_{I=1}^{N} m_I \dot{R}_I^2 \right\rangle_t}{f k_{\text{B}}} \tag{16.2}$$

The number N in Equation 16.2 gives the number of particles in the system, which is now one for the single CO molecule.

ergodic theorem

We will obtain the same value as we would get if we looked at the temperature of a snapshot of a simulation with a large number of CO molecules. This equivalence is called the ergodic theorem: the average of a quantity calculated for a snapshot of a large system equals the time average of the same quantity for an equivalent small system. For an ideal gas, this equivalence is indeed obvious. With the ergodic theorem, we now have the option to run an MD trajectory, monitoring the instantaneous value of a snapshot temperature as well as the temperature average over the simulation.

We need to keep in mind that we have to be careful when allowing external degrees of freedom in the simulation by adding a momentum and an angular momentum to the system, and consequently adjust $f = 3N$. However, velocity contributions that can be assigned to external degrees of freedom do not contribute to internal processes of the system, for example, to diffusion, to chemical reactions, or to significant rearrangements such as protein folding. In practice, one usually keeps momentum and angular momentum zero.

16.2.3
The Connection to Real Macroscopic Systems

The MD simulation of carbon monoxide, carried out as above, does not reflect a realistic, statistically correct behavior of the system. If we reduce the simulation to a single particle, we also reduce the total energy to that of a single particle. Within the ideal-gas approximation, the velocity distribution is described by the Maxwell–Boltzmann distribution (P denoting the probability, m the mass of the particle, and $\dot{R}^2$ is defined as above), **Maxwell–Boltzmann distribution**

$$P(\dot{R}) = \sqrt{\frac{2}{\pi}} \left(\frac{m}{k_B T} \right)^{3/2} \dot{R}^2 e^{-m \dot{R}^2 / 2 k_B T} \qquad (16.3)$$

A single CO molecule has only one degree of freedom and therefore cannot produce any velocity distribution. However, even if we record the velocities of the individual steps of the MD trajectory of a single CO molecule, the velocity distribution will be determined by the shape of the potential function. In practice, a more realistic behavior is obtained for larger system sizes, as they exhibit many more coupled degrees of freedom. If the system size is too small, one has to perform, for example, real gas simulations with a large number of particles, or couple the system with an appropriate external heat bath (see next section). For very large systems, the velocities would form a Maxwell–Boltzmann distribution at any snapshot.

We also have to bear in mind that MD simulations usually run over a time of several picoseconds. Chemical processes, on the other hand, typically happen on a time scale that is 6–10 orders of magnitude longer. A chemical process is connected to a specific situation regarding the participating molecules. For example, in a rotating methyl group, all velocity vectors of the hydrogen atoms point along the rotating direction. For such an event, there is only a specific probability, and a simulation must last long enough in order to observe it. There are several ways to increase the probability of an event, the simplest one being to increase the simulation temperature and thus the kinetic energy of the particles, because then activation barriers may be overcome.

The reduction of time scales of chemical reactions by increasing the simulation temperature is consistent with real chemistry in the test tube.

16.2.4
External Heat Baths – Thermostats

thermostat

Often we want to simulate a system at a certain, previously defined, temperature. This can be realized by a thermostat, that is, the simulation of a coupling macroscopic heat bath. The thermostat simulates a canonical ensemble, as it allows heat transfer to and from the system. The process of setting the system to a certain temperature is called equilibration, and to equilibrate a system we use the canonical ensemble (*NVT*) and perform an MD simulation until the averaged temperature remains constant within numerical limits for reasonable averaging intervals. The equilibration time depends strongly on the initial starting structure of the system. Therefore, equilibration is a critical and important part of the MD simulation that has to be performed thoroughly.

equilibration

Commonly, we speak of the *NVT* ensemble, even though the instantaneous temperature is usually not constant, but for a long time its average corresponds to a certain temperature. In this section we briefly outline the most common algorithms that can be used to maintain the simulation temperature. For a more detailed discussion, the reader is referred to the literature [1, 2].

16.2.4.1 A Simple Scaling Thermostat
For a reasonably large system, we can just set the instantaneous temperature by scaling the velocities by a simple factor. After determining the temperature using Equation 16.1, we scale all the velocities by

$$\dot{\boldsymbol{R}}_I \to \dot{\boldsymbol{R}}_I \sqrt{T_0/T} \tag{16.4}$$

simple scaling thermostat

T_0 being the desired simulation temperature. In this way, the temperature of the system is kept constant at all times. Unfortunately, our system is usually too small for this simple scaling thermostat to be applied to perform reasonable simulations. This will be illustrated in the Demonstration. The stubborn application of Equation 16.4 results in hazardous numerical issues for systems with few – in particular with a single – internal degrees of freedom.

16.2.4.2 The Berendsen Thermostat
It is more reasonable to adjust the instantaneous temperature in such a way that the time-averaged value approaches the desired temperature, as defined in Equation 16.2. The Berendsen thermostat is a simple way to carry out this procedure. Again, individual velocities will

Berendsen thermostat

be scaled to adjust the temperature, but with some moment of inertia:

$$\dot{\boldsymbol{R}}_I \rightarrow \lambda \dot{\boldsymbol{R}}_I, \quad \lambda = 1 + \frac{1}{2}\frac{\Delta t}{\tau}\left(\frac{T_0}{T} - 1\right) \tag{16.5}$$

The scaling factor λ thus depends on some coupling parameter τ and its relation to the MD time step Δt. Application of Equation 16.5 is almost as simple as that of the simple scaling thermostat in Equation 16.4, and its influence on the system can easily be controlled by the coupling parameter τ. Indeed, in practice the value of τ can be increased, and once the target temperature is reached over a reasonable period the thermostat can be switched off completely. Note that, if this thermostat is applied to the velocity Verlet algorithm and velocities are scaled at each half integration step, the half time step must be inserted in Equation 16.5.

16.2.4.3 The Andersen Thermostat

The scaling thermostats have the disadvantage that no additional momenta can be introduced into the system, and consequently the velocities of a (too small) system do not form a Maxwell–Boltzmann distribution, a requirement for the correct prediction of statistical processes. Assume, for example, that the oscillations of a water molecule can be perfectly described within the harmonic approximation. As we know from Chapter 12, the three internal vibrations are independent of each other. If we excite only one internal degree of freedom, for example the symmetric stretching mode, the system would – within an ideal simulation technique – remain in this mode forever. Collisions with an external bath would, however, also excite other degrees of freedom of the system. This can be realized conveniently by the Andersen thermostat. Here, the system is subject to collisions with fictitious particles of an external heat bath. Each individual atom is subject to an inelastic collision, depending on a probability that depends on a rate constant $0 < r < 1$ (e.g., for $r = 0.1$, each atom has a 10% probability of colliding with a fictitious particle of the heat bath). The particle's velocities after the collision form the Maxwell–Boltzmann distribution of the desired temperature (Equation 16.3). This simple thermostat is easy to implement. If applied to Cartesian coordinates, it introduces stochastic momenta and angular momenta into the system, which have to be accounted for by setting $f = 3N$.

Andersen thermostat

16.2.4.4 The Nosé-Hoover Thermostat

All thermostats have in common that they transfer heat, sometimes even momenta and angular momenta, to or from the system. Obviously, the total energy then loses its property as a control parameter of the system, and it is no longer obvious whether our simulation is on safe numerical grounds (see Chapter 15). Nosé and Hoover formulated a method, the Nosé–Hoover thermostat, where the energy accounting of

Nosé–Hoover thermostat

the heat bath is realized by one (or more than one) fictitious particle, which is propagated like the real particles of the system. This particle acts as a reservoir of energy for the heat bath. If the Hamiltonian of the system is extended properly by this external particle, the instantaneous expectation value of the extended Hamiltonian, that is, the total energy of the extended system, remains constant, and again a control parameter is available.

16.2.5
Averages

The calculation of the temperature already shows the importance of calculating averages, as instantaneous information is not useful for the comparison with experiment. Indeed, the calculation of averages is more involved than it appears at the beginning. Assume that we want to calculate the average intermolecular distance of the carbon monoxide molecule. Within the harmonic approximation, for obvious reasons, we can report the result as the equilibrium distance. For a more accurate potential, for example, the Morse potential, inspection of Figure 15.1 tells us that the molecule spends more time at distances that are longer than the equilibrium distance than at distances that are shorter than the equilibrium distance. The increase of the time-averaged bond length with temperature is well known from X-ray spectroscopy. Note that averages have to be taken carefully. For illustration, assume that your computation time is too short, and that you take the average only over a certain time period that includes – by chance – one half-period in the long-distance part of the oscillation (see Figure 16.1). In this way we overestimate the bond length significantly due to the artifact that the simulation time is too short and the selection of points to calculate the average in Equation 16.2 is not well chosen. If we increase the simulation time, the artifact disappears linearly in time.

To avoid artifacts like this in the first place, there are two strategies. First, we could select a random number of snapshots and control whether the result converges against a certain value. A second possibility would be to calculate the time average of the same trajectory, but varying the starting point, so that averages of different simulation period are compared. A detailed discussion on this important point can be found in [3].

16.3
Demonstration

We start this demonstration by extending the spreadsheet for the velocity Verlet algorithm for the vibrational motion of carbon monoxide

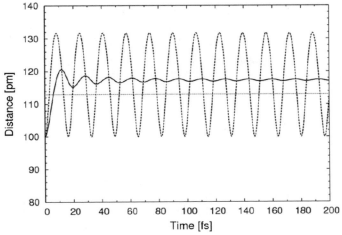

Figure 16.1 Distance (dashed line) and average distance (solid line) of the CO molecule as functions of simulation time obtained from the MD simulation in Chapter 15. The horizontal dotted line shows the equilibrium distance of the CO molecule (113 pm). Note that the average is higher than the equilibrium bond length due to the asymmetric Morse potential. Averaging over only one half-period with $R > R_e$ results in an average of 125 pm; for the half-period with $R < R_e$ the average is 105 pm (for a harmonic potential, these two averages are equal).

that we programmed in the Demonstration of Chapter 15. Save the spreadsheet as a new file. We see immediately that the velocity Verlet algorithm has the advantage that we know the velocities at each iteration. Open a new column to convert the kinetic energy, which we have already calculated at each step n, into the temperature. Use Equation 16.1 for this purpose. As we have used molar units (energies in kJ mol^{-1}), we make use of the universal gas constant R, the product of Boltzmann's constant and Avogadro's number, $R = k_B N_A = 8.3144$ J mol^{-1} K^{-1}, and we obtain the simple expression $T = 2E_{kin}/(fR)$, with $f = 1$.

If we start again with an initial distance of 100 pm, we observe that the temperatures reach enormous values of more than 30 000 K! Obviously, molecular vibrations at ambient conditions have much smaller amplitudes. However, let us have a look at some mean values of this computer simulation. For a time step of 0.1 fs, an oscillation takes approximately 164 steps, corresponding to a period of 16.4 fs. During this oscillation, the system starts with zero velocity ($T = 0$ K), passes through the potential minimum with a velocity corresponding to a temperature of nearly 32 000 K, drops back to zero for the long-distance extremum, passes through the potential minimum (32 000 K), and eventually drops back to the initial temperature and distance after about 16.4 fs. The average temperature over the period of 164 simulation steps is

$$\langle T \rangle_{t=16.4\text{ fs}} = \frac{1}{164} \sum_{i=1}^{164} T^{(i)} = 15\,245\text{ K}$$

which we would consider to be the simulation temperature. For a perfectly harmonic oscillation, the average bond length is the equilibrium distance, 113 pm. The average bond length of this computer simulation is larger,

$$\langle R \rangle_{t=16.4\text{ fs}} = \frac{1}{n} \sum_{i=1}^{n} R^{(i)} = 117.02\text{ pm}$$

This results from the fact that the molecule has larger oscillations to longer distances, as the potential is shallower in this region, and is reflected by the shortest distance during the simulation (100 pm, 13 pm less than the equilibrium distance) and the longest distance (131.7 pm, 18.7 pm more than the equilibrium distance).

Let us now reduce the initial bond length to 110 pm. We obtain a simulation temperature of 665.1 K, a value that might be used for the simulation of realistic systems in the gas phase. The average bond length is now 113.17 pm, very close to the equilibrium position, and the molecule oscillates between 110 pm and 116.22 pm. Also the vibrational frequency has changed: one period lasts now 15.4 fs, not 16.4 fs as for the strong oscillation. This value corresponds to a value of $\omega_0 = 4.08 \times 10^{14}\text{ s}^{-1}$, which is very close to the value obtained using the harmonic approximation in Chapter 3. Obviously, the harmonic approximation holds only for small oscillations.

So far, we have modified the simulation temperature by setting the initial condition, that is, the starting value for the interatomic distance and the velocity. Now, let us try to set the temperature by a thermostat. For a system with only one degree of freedom, the simple scaling thermostat has fatal numerical problems, as application of Equation 16.4 requires division by zero for iterations where $T = 0$ K. In realistic systems, the probability for such situations will be negligible, except for situations with very special starting conditions. We could avoid this dilemma for our system by always setting the velocity to a constant value that corresponds to the desired simulation temperature. Obviously, this is not a solution, as the molecule will always end up in unphysical situations: it would either eventually dissociate (for starting distances shorter than the equilibrium distance) or undergo "nuclear fission" for longer distances.

The Berendsen thermostat, applied as in Equation 16.5, also has the problem that a division by zero is required if the instantaneous temperature is zero. We can avoid this problem by putting Equation 16.5 onto safe numerical grounds. We introduce a slight modification that avoids the division by zero:

$$\dot{\boldsymbol{R}}_I \rightarrow \lambda \, \dot{\boldsymbol{R}}_I, \quad \lambda = 1 + \frac{1}{2} \frac{\Delta t}{\tau} \left(\frac{T_0}{\max(1 \text{ K}, T)} - 1 \right) \tag{16.6}$$

Let us now figure out how a thermostat acts. We change the spreadsheet, so that it runs over a long simulation period, say, 10 000 steps with a time step of 2.5 fs, which corresponds to a total simulation time of 25 ps. We program Equation 16.6 using the instantaneous temperature, and request the temperature of the heat bath to be 300 K. We activate the thermostat by adding a new column, which includes the modified velocity as given in Equation 16.6, and use this velocity for the estimation of the half-step velocity (Equation 15.9).

First, we switch the thermostat off by setting $\tau = 1000$ ps, a very large value for which Equation 16.6 takes no effect. For the entire 25 ps we obtain an average temperature of 663 K. Now we activate the thermostat, by setting the coupling constant to a typical value, for example, 0.5 ps. The simulation temperature drops to 431 K. It is interesting to calculate the temperature in succeeding simulation intervals. During the first 5 ps, the average temperature is 586 K, for the next 5 ps it drops to 472 K, and it continues to decrease. Note that the target value of 300 K is not reached even after 25 ps, the average temperature is still 334 K! We can speed up the equilibration by reducing the coupling constant, for example, to $\tau = 0.1$ ps. In this case, we reduce the temperature faster, and after about 10 ps the average temperature converged to a value of 294 K. The Berendsen thermostat will never reach the target value with this short coupling constant, but we can achieve a pre-equilibration in much quicker time. In practice, it is advisable to increase the coupling constant successively. Production simulations should have values of $\tau \geq 0.5$ ps.

Now we will turn to a problem with more than one degree of freedom, the water molecule. This part of the Demonstration is written for deMon and Molden. If you want to use a different computer code, you may need to change some details here, but the general strategy will remain the same.

First, we need to optimize the water molecule at the computational level at which we will carry out the molecular dynamics simulation. For this purpose, we generate a Z-matrix using Molden. The input for the geometry optimization should look like:

```
TITLE WATER OPTIMISATION
VXCTYPE AUXIS PBE
BASIS (DZVP-GGA)
AUXIS (GEN-A2)
OPTIMISATION
GEOMETRY ZMATRIX
O
```

```
H 1 HO2
H 1 HO3 2 HOH3
VARIABLES
HO2 0.95
HO3 0.95
HOH3 104.471
CONSTANTS
END
```

Optimize the geometry and make sure that the geometry optimization has converged (check Chapter 9). For deMon with the input above you obtain values for HO2 and HO3 of 0.9773 Å and an angle HOH3 of 104.3965°. Use the equilibrium geometry as initial structure for the molecular dynamics simulation. For deMon, you can either use the .new file that has appeared in your folder after the optimization was complete, or create a new input file. Note that the computational level should remain the same.

We will first run a microcanonical molecular dynamics simulation (*NVE*) where the symmetric stretch vibration is excited. For this purpose, we give an input file with the equilibrium bond angle, but with slightly shorter bond length (0.93 Å). We also include the molecular dynamics keywords. deMon employs the velocity Verlet algorithm. We run 100 MD steps with a time step of 0.25 fs and write the trajectory output at each iteration (INT=1). The thermostat is switched off (BATH NONE) and initial velocities are zero. The keyword VISUALISATION MOLDEN MD gives the trajectory as a file that can be interfaced with the Molden program. We have to request a tight numerical tolerance to have a very good quality of energy gradients, which is ensured by the keyword SCFTYPE TOL=0.5E-7. Thus:

```
TITLE WATER MOLECULAR DYNAMICS
DYNAMICS MAX=100 STEP=0.25 INT=1
BATH NONE
VELOCITIES ZERO
VISUALISATION MOLDEN MD
VXCTYPE AUXIS PBE
BASIS (DZVP-GGA)
AUXIS (GEN-A2)
SCFTYPE TOL=0.5E-7
GEOMETRY ZMATRIX
O
H 1 HO2
H 1 HO3 2 HOH3
VARIABLES
HO2 0.93
HO3 0.93
```

```
HOH3 104.3965
CONSTANTS
END
```

After the simulation ends (it may run for a couple of minutes, depending on your computer), we inspect first the output file. The simulation was running for 25 fs. The period of the symmetric stretch mode is about 9 fs. During the whole simulation, the total energy remained nearly constant with a numerical oscillation of 0.04 mHartree. The average simulation temperature was 160.8 K. Note that this temperature has been calculated for all degrees of freedom, but we have activated only one. Now visualize the potential energy during the simulation and observe the molecular motion using `Molden`. If you click on the **Geom. Conv.** button, you see the oscillations of the potential energy. If you rotate the molecule into an orientation in which you can see the structure and click on **Movie**, you can see the oscillations and the corresponding potential energy. The sine-like appearance of the potential energy with respect to the simulation time indicates that this oscillation is – in very good approximation – a harmonic one.

As the next step we activate two vibrations, the symmetric stretch mode and the bending mode, by changing the angle to 96°. All the other parameters are identical to those in the simulation above, except that we need to run the simulation for 400 iterations (`MAX=400`). The `Molden` output now shows a more complicated function of the potential energy surface with simulation time, the combination of the bending and symmetric stretching modes. The bending mode has a longer period (i.e., a lower frequency), and we can also see that even the strong elongation of more than 7° from the equilibrium position results in a significantly smaller amplitude for the vibration of the bending mode. Also there is an increase in the average simulation temperature, which is now 255 K.

Finally, we start the simulation from the equilibrium position, give random velocities to the atoms, and employ a thermostat, the Nosé–Hoover thermostat, as briefly discussed above:

```
TITLE WATER MOLECULAR DYNAMICS
DYNAMICS MAX=400 STEP=0.25 INT=1
BATH NOSE T=300
VELOCITIES RANDOM
VISUALISATION MOLDEN MD
VXCTYPE AUXIS PBE
BASIS (DZVP-GGA)
AUXIS (GEN-A2)
SCFTYPE TOL=0.5E-7
GEOMETRY ZMATRIX
O
```

```
H 1 HO2
H 1 HO3 2 HOH3
VARIABLES
HO2 0.9773
HO3 0.9773
HOH3 104.3965
CONSTANTS
END
```

If you inspect the potential energy as a function of simulation time, you should realize that there is little structure – the system has its three degrees of freedom, and in addition it couples with an external system. This is the typical behavior of the potential energy during a molecular dynamics trajectory.

16.4
Problems

1. Molecular dynamics of F_2 in a spreadsheet

Run an MD trajectory of the fluorine molecule F_2 (for parameters, see Table 3.1) in a spreadsheet. Use the velocity Verlet algorithm and the Berendsen thermostat, as modified in Equation 16.6.

a) First, start with a bond length of 130 pm and switch the thermostat off. Simulate for 30 ps and determine the average temperature and the average intermolecular distance. Compare the latter value with the equilibrium distance.

b) Switch on the thermostat and equilibrate the molecule at 300 K with a tolerance of 3%. For this purpose, successively increase the coupling parameter from 0.1 over 0.5 and 1 ps until you finally run the molecule in the *NVE* ensemble. Take averages in intervals of 5 ps.

c) Investigate, using your spreadsheet of part 1a, when the harmonic approximation becomes inaccurate for F_2. For this purpose, increase the amplitude of the molecular vibration, run the trajectory in the *NVE* ensemble, and calculate the average intermolecular distance $\langle R \rangle$. Plot $\langle R \rangle$ as function of the simulation temperature $\langle T \rangle$ and of the amplitude $\frac{1}{2}(R_{max} - R_{min})$.

2. Molecular dynamics for methane using a computational chemistry computer code

Use computational chemistry software, for example, deMon as delivered with this book, to perform several different MD simulations of the methane molecule.

a) Perform a geometry optimization (see Technical Details in Chapter 9).

b) Perform an MD simulation with the excited symmetric stretch mode. Determine the vibrational frequency by monitoring the potential energy as a function of time and compare this value with a frequency analysis.

c) Run an MD simulation with two modes excited, for example, a stretching mode and a bending mode. Can you determine the two vibrational frequencies?

d) Start a trajectory using the Nosé–Hoover thermostat (see Demonstration) and discuss the behavior of the potential energy as a function of simulation time for methane. Why is it much more complex than that of water?

16.5

Review and Summary

In this chapter, we have deepened our knowledge about molecular dynamics simulations by introducing heat baths and thermostats. Therefore, we have briefly discussed the canonical ensemble (NVT) and the microcanonical ensemble (NVE) in general and in the context of a molecular dynamics simulation.

As a control parameter for a system coupled to a heat bath, we have introduced the temperature, which is directly connected to the kinetic energy of the system. Moreover, we have used the ergodic theorem in order to scale down the system size to a single molecule. It states that the average of a quantity calculated for a snapshot of a large system equals the time average of the same quantity for an equivalent small system.

We have also introduced different thermostats that can be used to equilibrate the system within an MD simulation: a simple scaling thermostat, which scales the instantaneous temperature; the Berendsen thermostat, which scales the average temperature; the Andersen thermostat, which randomly introduces additional momenta to the atoms; and the Nosé–Hoover thermostat, which propagates the energy as a fictitious particle and makes the total energy again available as a control parameter. During the discussion of these different thermostats, we have seen that appropriate averaging of the system's quantities is important.

In the first part of the Demonstration, we have extended the spreadsheet of the velocity Verlet algorithm from Chapter 15. Calculating the temperature has shown us that instantaneous values of more than 30 000 K occurred when starting from a bond length of 100 pm. The anharmonicity of the Morse potential could be observed by calculating average bond lengths over the two half-periods (below and above the equilibrium bond length). By reducing the starting bond length, the average temperature dropped and the average bond length came close to the equilibrium bond length. Applying the Berendsen thermostat to the problem made it possible to equilibrate the system to the desired temperature.

In the second part of the Demonstration, we used the quantum chemical code deMon to perform an MD simulation of the water molecule. Here we could see that it is possible to excite molecular vibrations directly. In an *NVE* ensemble the molecule kept vibrating in the excited mode. Exciting more than one mode showed that the superposition is also retained, but that the variation of the potential energy is less clearly interpretable. Introducing randomly chosen velocities excites all modes, and an interpretation of the potential energy curve is only possible with more advanced numerical tools (Fourier analysis).

References

1 Allen, P. and Tildesley, D.J. (1989) *Computer Simulation of Liquids,* Oxford University Press.

2 Frenkel, D. and Smit, B.J. (2002) *Understanding Molecular Simulation,* Academic Press.

3 Schatz, G.C. and Ratner, M.A. (2002) *Quantum Mechanics in Chemistry,* Dover Publications.

17
Molecular Dynamics – Simulated Annealing

17.1
Aim

In this chapter we will use molecular dynamics to explore the potential
energy surface of a molecule and to discuss global geometry optimization. We will use the simulated annealing approach together with
traditional geometry optimization techniques. We will combine what
we have learned in Chapters 9, 13, 15 and 16, and see how, for example,
the structure of a novel molecule can be found, and how different
isomers can be identified by exploring the potential energy surface
with computational chemistry tools.

17.2
Theoretical Background

In Chapter 9 we have discussed different algorithms that can be used
for the optimization of a molecular structure. All the optimization
routines that we have discussed so far, like the steepest descent method
or the Newton–Raphson algorithm, are suitable to find a local minimum,
a structure that is usually close to the chosen starting geometry. Usually,
this is exactly what a chemist wants: the starting geometry is built by
using chemical intuition, and the geometry optimization is essentially
employed to obtain quantitatively correct structural parameters such
as bond lengths and bond angles. For many stoichiometries various stable
isomers are known. Let us, for example, take the difluoroethene $(C_2H_2F_2)$ **isomer**
molecule. Spontaneously we can imagine three isomers as shown in
Figure 9.1. Their total energies, determined using the same details as
given in Chapter 9, are -276.8259 Hartree (1,1-difluoroethene),
-276.8106 Hartree (*cis*-difluoroethene), and -276.8099 Hartree
(*trans*-difluoroethene). Thus, the 1,1-difluoroethene isomer is the

Computational Chemistry Workbook: Learning Through Examples
Thomas Heine, Jan-Ole Joswig, and Achim Gelessus
Copyright © 2009 WILEY-VCH Verlag GmbH & Co. KGaA, Weinheim
ISBN: 978-3-527-32442-2

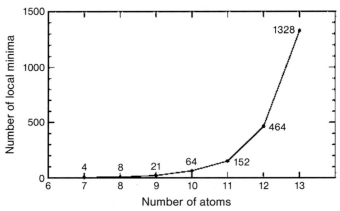

Figure 17.1 Number of local minima [2] on the potential energy surface of clusters described by a Lennard-Jones pair potential.

most stable within this computational method.[1] All three isomers are stable and can be isolated; and if the isomers could be converted easily into each other, all three isomers could be found at thermodynamic equilibrium at finite temperatures (see Chapter 14). The potential energy surface, that is, the energy barriers that need to be overcome when converting one isomer into another, determines the conversion rates. For covalent bonds, such conversions require high temperature, as we will see below.

If we want to investigate molecules or clusters that exhibit non-covalent bonds, for example, all-metal clusters, the energy barriers between isomers are low, and it is necessary to find the global minimum structure as it is always a significant fraction of the produced isomer mixture. The global minimum is always the reference if a molecule is created in a thermodynamic process, for example, for fullerenes, which are produced in an arc discharge chamber. For such structures, the bonding pattern is not at all obvious. Moreover, even for relatively small clusters, the number of possible isomers may become very large. This has been investigated by employing simple pair potentials – in fact, the number of local minima of a polyatomic cluster with respect to the number of atoms increases more strongly than exponential (Figure 17.1) [1]. The resulting problem of finding the lowest energy isomer (corresponding to the global minimum structure) is thus a so-called NP-hard problem.[2]

---

1) Note that the energy differences between the most stable and the other two isomers are only $40 \, kJ \, mol^{-1}$ and $42 \, kJ \, mol^{-1}$. Using a different functional or basis set might change the order of the isomers.

2) The term "NP-hard" stands for "non-deterministic polynomial-time hard" and the NP-hard problem is at least as hard as the hardest problems in non-deterministic polynomial time. An example is the "traveling salesman problem."

17.2.1
The Potential Energy Surface

As we have already discussed in Chapter 9, the potential energy surface of a molecule is usually a multi-dimensional function (cf. Figures 9.2 and 17.2). The Born–Oppenheimer approximation treats the nuclei classically, and the potential is therefore equivalent to the potential energy, and we use synonymously $V = V(R_1, R_2, \ldots, R_N) = E_{pot} = E_{pot}(R_1, R_2, \ldots, R_N)$. In the previous two chapters we have, moreover, seen that a system always has a certain amount of kinetic energy. Adding these two – that is, the potential arising from the interactions between the particles, and the kinetic energy resulting, for example, from a certain temperature – leaves us with the total energy of the system. In a molecular dynamics simulation within the *NVE* ensemble, we can reach only those parts of the potential energy surface that are below the total energy and that are not separated by barriers (see Figure 17.2).

Now we can again interpret temperature as kinetic energy and draw the analogy to experiment: temperature is the kinetic energy of the atoms of the system, and their motion is equivalent to shifting the molecule on the potential energy surface. Many chemical reactions require that the molecules pass through an energy barrier – in the laboratory we heat up the molecules for this purpose. Here, we can do exactly the same. We increase the temperature, which means that the molecules have more kinetic energy, the atoms can move faster,

potential energy surface

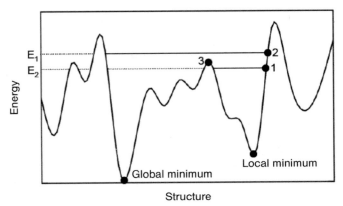

Figure 17.2 A fictitious potential energy surface. Starting point 1 has a total energy E_1, in this case entirely as potential energy, so starting velocities are zero. If the system approaches the nearest local minimum, part of the potential energy is converted into kinetic energy. The total energy at point 1, however, is not sufficient to overcome the barrier 3. Starting at point 2 on the PES gives the system enough energy to overcome this barrier and to reach the global minimum. This is equivalent to "heating" the system by increasing the temperature of the simulation; hence the kinetic energy increases, and so does the total energy.

and consequently they can overcome higher barriers. Increasing the temperature means that a larger area of the potential energy surface can be scanned in an MD simulation (see Figure 17.2).

In reality, however, overcoming a barrier means that sufficient kinetic energy is used for collective motion of part of a molecule, for example, in the rotation of a functional group. Let us illustrate this with a simple example. If *cis*-difluoroethene (*cis*-$C_2H_2F_2$; see Figure 9.1) is to be converted to *trans*-$C_2H_2F_2$, it is necessary that one CHF group rotates with a kinetic energy that is large enough to overcome the (quite large!) rotational barrier that breaks the double bond. The carbon atom is on the rotational axis and does not contribute to the kinetic process. Its electrons, however, strongly contribute to the change of the potential during this process. Hence, it requires that the F and H atoms of the CHF group rotate simultaneously. Even if the total energy of the system is just enough to overcome the barrier for the conversion of the two isomers, this event will have a very low probability. In an experiment, this means that we have a slow reaction. In a computer simulation, this means that we will most probably not observe this process, owing to the much shorter simulation times: the stochastic distribution of the velocities needs – by chance – to correspond at an instant solely to the rotational motion of the H and F atom with all other atoms at rest. The probability for the conversion obviously increases for higher temperatures: even if the other atoms of the molecule move, there is still a chance that the two atoms have enough energy for the rotation.

The approach to finding several local minima on the potential energy surface – if not even the global minimum – by heating a molecule or cluster in a computer simulation to explore the potential energy surface **simulated annealing** is called simulated annealing, and it is frequently used in computational cluster physics as a tool for global geometry optimizations.

17.2.2
Simulated Annealing

The simulated annealing approach [3] can be used to search for the **global minimum** global minimum of the potential energy surface. From an initial molecular structure, a geometry optimization usually results in the nearest stationary point on the potential energy surface. This is a local minimum (cf. Chapter 9). In many cases, the resulting structure has the same topology as the starting structure, for example, a steepest descent optimization of the HCN molecule never would result in a CNH arrangement.

In the simulated annealing approach, the total energy of the system will be increased successively. This usually happens if we start with a finite temperature from a local minimum. The system will move, hence the potential energy will increase, and if we request a high temperature, the system requires additional energy – it is heated. By increasing the total energy in this way, the system can overcome energy barriers. The longer we simulate, the larger the areas of the potential energy surface that are scanned. After the simulation, we inspect the potential energy as a function of computer time. We will find places where the potential energy has a local minimum (as a function of time – this is not a local minimum structure!). Now we take the geometries of these "snapshots" and "cool the structure down." This is most efficiently done if we just perform a geometry optimization of these structures, for example, with a steepest descent.

We will end up with a series of local minimum structures, and we can compare them, by their structures, their relative energies, and their properties (e.g., vibrational spectrum, ionization potential, etc.). In the Demonstration, we will follow an example of recent research. The global minimum of the all-metal cluster Al_4^{2-} is a square, planar aromatic molecule that has been detected in the gas phase and characterized by its vertical detachment energy (see Chapter 13) [4]. We will determine its structure by the simulated annealing procedure. A more involved problem is treated in problem 2: the all-metal cluster Li_6 has several local minima (chain, ring, pyramid, etc.). Particularly stable local minima are all indicated in a molecular dynamics simulation at high temperature. After the topological structures of the minima have been identified, low-potential snapshot structures are optimized and the isomers can be identified. If the potential energy surface is scanned sufficiently well, the lowest energy structure will be the global minimum. There is, however, no guarantee that there is no isomer of lower energy.

Finally, it should be noted that the level of approximation of the computational method to compute the potential energy surface does not necessarily need to be of highest accuracy – the obtained isomers will be refined by individual geometry optimizations, which have to be performed with a method ensuring the requested accuracy. Therefore, we will use a more approximate approach to solve the Schrödinger equation than usual for the molecular dynamics part of the simulated annealing approach.

17.3
Demonstration

17.3.1
Inspecting Stationary Points With Simulated Annealing

In Chapter 11 we learned that a stationary point of the potential energy surface may correspond not to a minimum, but, for example, to a

transition state transition state. Traditionally, vibrational spectroscopy is used to characterize a minimum in computational chemistry. If all the frequencies of the molecule are real, positive numbers, and if the energy gradients are zero, then we have found a local minimum. However, this approach is often not practicable, as frequency analyses become very computer time expensive with increasing system size. An alternative is simulated annealing.

As a simple example, construct the planar D_{3h} transtition state of an ammonia molecule. We deliberately choose a wrong symmetry here. You can use the Z-matrix editor of `Molden`. The resulting Z-matrix – prepared for the `deMon` input – may have the form:

```
GEOMETRY ZMATRIX
N
H   1 HN2
H   1 HN3   2 HNH3
H   1 HN4   2 HNH4   3 DIH4
VARIABLES
HN2   1.008
HN3   1.008
HNH3  120.0
HN4   1.008
HNH4  120.0
DIH4  180.0
CONSTANTS
END
```

The geometry optimization will not change the molecular symmetry, because there is no gradient pointing out of the molecular plane and no atom experiences a force that drags it out of that plane. After a few iterations the stationary point is found, but the molecule is still planar, and only the bond lengths are adjusted. A frequency analysis would reveal an imaginary frequency, corresponding to an out-of-plane mode (you can check that by adding the optimized geometry and the keyword FREQUENCY to the input file and rerun the calculation). Simulated annealing immediately indicates that the stationary point does not correspond to a local minimum. We start a molecular dynamics simulation from the optimized structure. For the `deMon` code the input looks as follows:

```
DYNAMICS MAX=400 STEP=0.5 INT=4
BATH NOSE T=600
VELOCITIES RANDOM T=1200 LP=0
VISUALISATION MOLDEN MD
BASIS (DZVP-GGA)
```

```
AUXIS (GEN-A2)
VXCTYPE AUXIS PBE
GEOMETRY Z-MATRIX ANGSTROM
N
H   1 HN2
H   1 HN3   2 HNH3           RAD
H   1 HN4   2 HNH4   3 DIH4  RAD
#
VARIABLES
HN2   1.013175
HN3   1.013175
HN4   1.013175
HNH3  120.000000
HNH4  120.000000
DIH4  180.000000
#
CONSTANTS
END
```

After the simulation, inspect the trajectory with the `Molden` editor. Obviously, the ammonia molecule immediately changes its structure to the well-known pyramidal form. Any snapshot structure (except the planar one) can now be selected and optimized, and the correct minimum will be found. For this purpose, you can select any structure in your molecular viewer. In `Molden`, it is convenient to activate the button **Geom. Conv.** and to pick a low-potential structure in the potential plot by clicking on the curve (opening the **ZMAT Editor** will give you the Z-matrix, which you can save). You can construct a geometry optimization input, and obtain the local minimum structure by performing the geometry optimization.

If you inspect the molecular dynamics trajectory more closely, you may notice that the ammonia molecule flips between two minima, crossing the planar transition state. This is the simulation of an atomic rearrangement, even though a very simple one.

17.3.2
Finding the Global Minimum of Al_4^{2-} Using Simulated Annealing

Exploring new molecular structures is exciting for any chemist. It is even more exciting if such a task can be performed when experimentalists and theorists cooperate closely. This has been done by the teams of Boldyrev, Wang, and Kuznetzov [4]. Particles of the stoichiometry $NaAl_4^-$ have been detected in the mass spectrum, and their vertical detachment energy has been measured. Further studies have shown that Al_4^{2-} is a stable double anion. Its structure was, however, not clear. There are several ways to arrange four Al atoms to form a metal cluster. Let us start from an obviously wrong one, a linear chain. We place four Al atoms with distance of 2.5 Å in a row and perform a geometry optimization. Remember that you calculate a dianion, so in deMon you need to set the keyword CHARGE −2.0. At the level as discussed in Chapter 9 the resulting geometry has the form:

```
AL   0.00   0.00    3.912
AL   0.00   0.00    1.283
AL   0.00   0.00   -1.283
AL   0.00   0.00   -3.912
```

We now perform a molecular dynamics simulation. We chose a slightly lower computational level (by selecting a smaller basis set, and by changing some technical details to converge the electronic structure) to reduce computer time in this example. If you have a powerful computer you can keep the computation at the higher level. Note that here we choose a relatively large time step, even though the temperature is quite high (1000 K). This choice has two motivations: first, the cluster contains only relatively heavy atoms (Al), and second, we do not need a high-quality trajectory.

```
DYNAMICS MAX=1000 STEP=2.0 INT=5
BATH NOSE T=1000
VELOCITIES RANDOM T=1500 LP=0
VISUALISATION MOLDEN MD
VXCTYPE AUXIS PBE
BASIS (A-VDZ)
AUXIS (GEN-A2)
CHARGE -2.0
SMEAR 0.05
DIIS OFF
GEOMETRY CARTESIAN
AL   0.00   0.00    3.912
AL   0.00   0.00    1.283
AL   0.00   0.00   -1.283
AL   0.00   0.00   -3.912
END
```

The trajectory shows that the linear isomer is quickly converted to a ring structure. Two isomers appear: a planar ring isomer, and a twisted isomer. A geometry optimization of those structures, however, quickly shows that the ring isomer is the more stable structure, and for a four-atomic molecule we are sure that it corresponds to the global minimum.

17.4
Problems

1. Identifying a stationary point that is not a minimum

Construct the ethane molecule in its eclipsed D_{3h} geometry. Perform a geometry optimization. Most optimizers will retain the symmetry. Check if the optimized structure is a local minimum by using the simulated annealing technique as discussed for ammonia in the Demonstration. Optimize a low-energy snapshot structure of the trajectory. Do you observe the correct D_{3d} structure?

2. Simulated annealing of Li_6

Construct different isomers of the all-metal cluster Li_6. Start from two initial structures: a linear chain, and a square, bipyramidal structure (see Figure 17.3). Optimize the starting structures. Then, start the simulated annealing. Optimize all reasonable isomers that occur in the resulting trajectory. Which isomers can you identify?

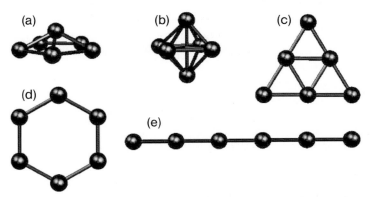

Figure 17.3 Five local minimum structures of the Li_6 cluster: (a) a pentagonal pyramid, (b) an square bipyramidal, (c) a planar structure, (d) a ring, and (e) a linear chain.

Hints

If the computations are not fast enough, use the even more reduced computational level:

```
VXCTYPE AUXIS VWN
BASIS (A-VDZ)
AUXIS (GEN-A2)
```

To avoid technical problems in solving the Schrödinger equation, use the additional keywords:

```
SMEAR 0.05
DIIS OFF
```

Some local minimum structures of Li_6 are given in Figure 17.3. The relative energy of these structures will depend on the computational level you use in your calculation. If you do not encounter these isomers, try to run longer trajectories and try to increase the temperature in the line:

```
BATH NOSE T=1000
```

17.5
Review and Summary

In this chapter, we have used the simulated annealing approach in order to explore the potential energy surface of a molecular structure. This approach makes use of the fact that the kinetic energy of a system enables it to overcome energy barriers on the potential energy surface. We can therefore use simulated annealing to find local minima that are not reachable by classical geometry optimization tools from the starting structure.

During a MD simulation of the system of interest at high temperature, the system will adopt different geometries with topologies that correspond to different local minimum structures. The molecule will stay in those topologies for certain time periods; nevertheless, it will vibrate. The observed structures can be selected and optimized separately. The resulting local minimum structures can then be compared and – if the potential energy scan is complete – that with the lowest potential energy is the global minimum structure. Moreover, the simulated annealing approach can be used to move the structure from a stationary point, for example, a transition state. In summary, the simulated annealing approach is a complementary tool to the approach to determine molecular structures by geometry optimizations from arbitrarily constructed starting structures.

In the Demonstration and the Problems sections, we have used this approach to find the global minimum for Al_4^{2-} and Li_6, and we have removed the NH_3 and the ethane molecules from their transition states. We have noticed that the level of calculating the electronic structure during the MD simulation might be decreased. A high level of approximation, for example, larger basis sets, should be used for the final geometry optimizations to compare the resulting structures.

References

1 Wales, D.J. (2003) *Energy Landscapes*, Cambridge University Press.

2 Tsai, C.J. and Jordan, K.D. (1993) *The Journal of Physical Chemistry*, **97**, 11227.

3 Kirkpatrick, S., Gelatt, C.D. and Vecchi, M.P. (1983) Optimization by simulated annealing. *Science*, **220**, 671.

4 Li, X., Kuznetsov, A.E., Zhang, H.F., Boldyrev, A.I. and Wang, L.S. (2001) *Science*, **291**, 859.

Appendix
The Computational Chemistry Software
Delivered with This Book

The *Computational Chemistry Workbook* builds on the learning-by-doing concept and requires the reader to solve problems in each chapter. Many of the problems require certain computational chemistry tools. The toolbox of a computational chemist gets more and more sophisticated year by year, and the beginner is confronted with the dilemma to choose an appropriate tool. We do not intend to solve this problem for each individual chemist, as too many factors must be considered when making this decision. However, we do want the reader to have a toolbox readily available that we consider to be suitable for solving the problems of the present workbook, and which works in most situations, such as on different hardware and in various software environments. Our choice is biased by the fact that the operating system of choice for professional computational chemistry is still Unix – today usually in the form of Linux. We are convinced that the present status of the Linux development is advanced enough to make this step, even if we are addressing people with limited interest in administrating computer systems (as some of the authors).

Linux is nicely suited for this workbook project as it allows the free distribution of a powerful software toolbox that was found to be sufficient for our purpose and, moreover, that we could extend and adopt to our needs. The software compilation on the accompanying CD is therefore distributed under the GNU Public License, which allows – among other things – the free redistribution of this software, and the adaptation of its content to other situations. However, note that some programs on the CD, in particular deMon, CaGe and Molden, have their own license agreements. If these programs are modified, then permission has to be granted by the respective authors.

We have chosen the Knoppix Live system as the basis for our project. Knoppix Live (http://www.knoppix.net) is a Linux system that is designed to run "out of the box" on almost all hardware based on the x86 design, which today includes all standard PCs, as well as newer-generation Apple Macintosh computers. Knoppix does not require a local Linux installation. The Linux operating system and all other

Computational Chemistry Workbook: Learning Through Examples
Thomas Heine, Jan-Ole Joswig, and Achim Gelessus
Copyright © 2009 WILEY-VCH Verlag GmbH & Co. KGaA, Weinheim
ISBN: 978-3-527-32442-2

software is directly loaded from the CD, and the standard configuration is a completely preconfigured working environment. This `Knoppix` system does not affect your computer's software installation on the hard drive. After you have removed the CD from the drive, your computer will work as usual. This makes it possible to use the CD in any computer laboratory where usually a different native operating system is employed.

Up-to-date information on the software for this workbook can be found at the project web page: http://www.compchen.jacobs-university. de/workbook.html.

A.1
Getting Started

There are – at least – three possibilities to run the `Knoppix` CD: booting from it, running it as a virtual machine, or running it from a USB stick. All three possibilities have their advantages and disadvantages, which we will highlight below. Alternatively, the computational chemistry software can be copied to almost any x86 `Linux` installation and used natively.

A.1.1
Booting Directly the Live System

Insert the CD into your computer's disk drive and boot your computer. You may have to change your BIOS settings to allow the computer to boot from CD, and make sure that you boot from CD and not from your local hard disk or any other device. This possibility is often the most convenient: you insert the disk, you reboot your computer, and the system is up. You may save data either by inserting a USB stick (it should be recognized as you insert it), or through the Internet. This option has two disadvantages: (i) starting applications is – at least for the first time – a bit slow, as they start from the CD, and (ii) you will have to reboot your computer whenever you want to use the software. Note the present `knoppix` version does not recognize most wireless network.

A.1.2
Booting as Virtual Machine

Copy the ISO image of the CD to your hard drive and set up the virtual machine, for example, `VMWARE`. Here is one of the many possible straightforward installation instructions that work on Windows Vista and XP, and similar solutions are certainly available for other operating systems.

1. Download and install the VMWARE player software, available at http://www.vmware.com/download/player/.

2. Download and install VMXWizard, available at http://rhysgoodwin. orconhosting.net.nz/vmxwizard/.

3. Copy the ISO image of the CD to your hard drive.

4. Set up the virtual machine using VMXWizard. Select

   a) Choose Linux 2.4 kernel and the necessary memory (use 256 MByte; for computers with 1 GByte memory and more, use 512 MByte).
   b) Ignore the hard disk (select 0).
   c) Choose the copied ISO image on your hard drive as DVD-ROM.
   d) Accept the suggestion for the Ethernet network.
   e) Label your Virtual Machine (e.g., Computational Chemistry Knoppix).
   f) Press **Next** and you are ready.

5. Start VMWare player and boot the virtual machine.

In this way you can start the software whenever you like, you can hibernate it if not needed, and you don't have to have the CD with you. The only drawback is that the computations may last a bit longer. Modern computers have features for fast virtualization; these features may have to be activated in the BIOS.

A.1.3
Running From a Pen Drive (USB Stick)

It is possible to run the software from a USB stick. This has the same advantages and disadvantages as if you run it directly from CD, except for three details.

1. You may run it from the USB stick on different computers without disk drive (laptop, netbook, etc.).
2. The execution of the programs might be faster due to the faster access of memory to the stick.
3. The Pen drive can be used to store data.

The preparation of the USB stick is straightforward, but a bit technical. The following instruction creates a bootable USB-memory stick from the Linux-Live CD. Two partitions are created on the USB-memory stick. The first partition has 750 MB and is used for the Linux installation. The second partition can be used to read and write user data.

1. Boot your computer with the CD and insert the USB-memory stick into a free USB port at your computer.

2. Open a terminal and type *su*

3. Type *fdisk –l* and note which drive is your USB-memory stick. Throughout this help page the letter x is used for the USB-memory stick. The letter x should be replaced by the actual letter for the USB-memory stick. For example, if the USB-memory stick is sdb, the letter x should be replaced by b.

4. Type *umount /dev/sdx1*

5. Type *fdisk /dev/sdx*
   5.1 Type *p* to show the existing partitions and *d* to delete it.
   5.2 Type *p* again to show any remaining partitions (if partitions exist, repeat the previous step).
   5.3 Type *n* to make a new partition.
   5.4 Type *p* for primary partition.
   5.5 Type *1* to make this partition number one.
   5.6 Hit enter to use the default first cylinder.
   5.7 Type *+750M* to make the first partition 750 MB large.
   5.8 Type *a* to make this partition active.
   5.9 Type *1* to select partition one.
   5.10 Type *t* to change its file system.
   5.11 Type *6* to select the fat16 file system.
   5.12 Type *n* to make another new partition.
   5.13 Type *p* for primary partition.
   5.14 Type *2* to make this the second partition.
   5.15 Hit enter to use the default first cylinder.
   5.16 Hit enter again to use the default last cylinder.
   5.17 Type *w* to write the new partition table.

6. Type *umount /dev/sdx1* to ensure the partition is unmounted.

7. Type *mkfs.vfat –F 16 –n usb /dev/sdx1* to format the first partition.

8. Type *umount /dev/sdx2* to ensure the partition is unmounted.

9. Type *mkfs.ext2 –b 4096 –L Workbook-data /dev/sdx2* to format the second partition.

10. Remove and reinsert your USB-memory stick.

11. Type *mount /dev/sdx1 /tmp/usb*

12. Type *cd /cdrom*

13. Type *cp –rf KNOPPIX boot/isolinux/* /tmp/usb*

14. Type *cd /tmp/usb*

15. Type *mv isolinux.cfg syslinux.cfg*

16. Type *cd*

17. Type *umount /tmp/usb*

18. Type *syslinux –sf /dev/sdx1*

19. Type *mount /dev/sdx1 /tmp/usb*

20. Type *cp /cdrom/index.html /tmp/usb*

21. Type *umount /tmp/usb*

22. Reboot the computer and set the system BIOS or boot menu to boot from USB device.

23. If booting fails it it possible that the MBR needs to be installed. In this case restart your computer with the CD and insert the USB-memory stick and then type the command *dd if=/usr/lib/syslinux/mbr.bin of=/dev/sdx*

The USB-memory stick also uses the ramdisk for the user data and newly created files or modifications to files are lost after a shutdown. The following commands change the location of the user data to the second partition of the USB-memory stick thus making file creations and modifications permanent.

1. Boot the computer with USB-memory stick
2. Open up a terminal and type *su*
3. Type *rm –f /home*
4. Type *mkdir /home*
5. Type *mount /dev/sdx2 /home*
6. Type *exit*

A.1.4
Configure Your Environment

After completion of the boot process, the file manager `Konqueror` automatically opens a window providing several links to software and documentation. At the lower margin of the computer desktop you find the control panel of the desktop environment KDE. The icons in the control panel provide quick access to frequently used files and applications. A mouse click on the "House" icon opens a new `Konqueror` window showing your `home` directory (your personal data). The subdirectory `examples` inside your `home` directory contains several example files for the computational chemistry software. The "Monitor" icon provides access to a terminal window. A terminal is an interface to the `Linux` operating system and is needed for all command-driven interactions with your computer. A short introduction to the most important `Linux` commands is given in the next section. The "Globe with rack-wheel" icon starts a window of the file viewer `Konqueror`.

The keyboard layout can be changed via the country flag on the right-hand side of the control panel. As default, the US keyboard layout is used. With a left mouse click on the country flag, the German or French keyboard layout can be selected. Many other country keyboard layouts can be easily added with a right mouse click on the country flag via the option **Configure**.

The Knoppix Live system does not utilize your computer's hard disk. All the files from your home account are stored in the computer RAM memory. After a shutdown or restart of the Knoppix Live system, all data are lost from the RAM. If you want to keep your data for later usage, they can be copied to the external USB device. The USB device is detected automatically by the Knoppix Live system upon insertion.

A.2
A Brief Introduction to Linux

Traditionally, Linux and other Unix versions use a command-driven interface for the interaction between user and operating system. With the emergence of affordable graphic hardware, the development of desktop environments (KDE, Gnome, CDE, etc.) was initiated, and nowadays many tasks can be performed easily by mouse clicks. Nevertheless, it still makes sense to work with commands, because they provide a greater flexibility and they can be used to write scripts for recurring tasks. The command line interface software to the operating system is called a "shell". Many different shells have evolved over the last decades, and currently the most frequently used shells are the bash (born again shell) and the tcsh (tc shell). The Knoppix Live system uses the bash.

Linux commands are typed behind the prompt in the shell, and upon pressing the **Return** key the command is executed and the result (if any) is written back to the terminal. Linux commands are case sensitive and for standard commands lower-case letters are always used. The behavior of most Linux commands can be modified with so-called option flags, which are added to the basic command after a space character. For example, the command ls lists files in the current directory, and the command ls - l lists files including their attributes (the option flag -l stands for "long listing"). Often, command options can be combined; the command ls - ldt shows files including their attributes (option l), keeps directories closed (option d), and sorts the files by modification time (option t). These Linux commands are summarized in Table A.1.

An easily available source of information about Linux commands and their options are the manual pages. The command man ls displays the manual page for the ls command. You can scroll up and down the

Table A.1 Using command flags for the example of the `ls` command.

Command	Description
`ls`	Show files in working directory
`ls -l`	Show files including file attributes
`ls -a`	Show also hidden files (files starting with " . ")
`ls -t`	Show files sorted by modification time
`ls -ldt`	Combination of options `l`, `d` and `t`
	Many, many more options

pages and leave them by typing `q` (quit). Another very useful source of information about `Linux`, `Linux` commands, and related topics is the Internet. Frequently used commands are shown in Table A.2.

Table A.2 Frequently used `Linux` commands.

Command	Description	Example
`cp`	Copy file	`cp file1 file2`
`mv`	Move (rename) file	`mv file1 file2`
`rm`	Remove file	`rm file`
`mkdir`	Create (make) new directory	`mkdir newdir`
`cd`	Change directory	`cd newdir`
		`cd ...`
		`cd`
`rmdir`	Remove empty directory	`rmdir newdir`
`pwd`	Print working directory	`pwd`
	("print" here means: show in the terminal)	
`man`	Show manual page	`man ls`
`date`	Show date and time	`date`

`Linux` allows the usage of wildcards to address groups of files instead of individual files only. Frequently used wildcards are listed in Table A.3. Examples of the usage of some of these wildcards are shown in Table A.4.

Table A.3 Wildcards in `Linux`.

Wildcard	Description
`?`	Exactly one arbitrary character
`*`	Any number (including 0) of arbitrary characters
`[abc]`	One matching character for that position
`[a-c]`	One matching character for that position
`~`	User's home directory

Table A.4 Some wildcard usage examples.

Command	Description
rm c*	Removes all files beginning with c
ls ab?	List all files beginning with ab having an arbitrary character at position 3
ls [ab,e-h]*	List all files beginning with a, b, e, f, g, or h
ls ~	List the content of your home directory
ls ~user7	List the content of the home directory of user7

In Linux, each file possesses attributes, which control read access (who may read a file), write access (who may create or modify a file), and execution permission (who may execute an executable file). Files are assigned to a user and a group, and the permissions for read, write, and execute can be set for the user, the group, and others (the rest of the world). The ls - l command shows the file attributes, for example:

```
-rw-r--r-- 1 knoppix knoppix 52 Jul 18 10:34 benzene
```

In this case, the file benzene belongs to the user knoppix and the group knoppix. File permissions are indicated by the 10 leading characters. The first position (in the example, not set) indicates special file types. Positions 2 to 4 indicate the read, write, and execute permissions for the user; positions 5 to 7 indicate read, write, and execute permissions for the group; and positions 8 to 10 indicate read, write, and execute permissions for others. In our example, the user has read and write permissions for the file benzene, while the group and others have only read permission. File permissions can be changed with the command **chmod**.

The commands in Table A.5 are helpful for file handling. It is recommended to practice these commands and to see how they behave and what output they produce. More details for each command can be obtained from each command's manual page.

The subdirectory examples in the user's home directory on the Knoppix Live DVD contains several files that can be used as test cases for the file handling commands. Start a shell by clicking the "Monitor" icon from the control panel and change to the examples subdirectory with the command cd examples. Then execute the commands listed in Table A.6 and try to understand the result.

In the examples above, the output is written to the terminal. Very often it is more useful to send the output to a file for further processing. This can be achieved with redirections. The command

```
cat h2.inp > output
```

Table A.5 Some helpful `Linux` commands.

Command	Description
`file`	Determine file type
`cat`	Show and concatenate files
`more`	Show file contents
`less`	Show file contents
`head`	Show first lines of a file
`tail`	Show last lines of a file
`grep`	Print all lines with a given pattern
`sort`	Sort lines of a file
`uniq`	Report or omit repeated lines
`find`	Search for files
`diff`	Compare files

Table A.6 `Linux` commands to be tested in the subdirectory `examples`.

Command
`file *`
`cat h2.inp`
`cat h2.inp h2.inp`
`more h2.inp`
`head - 5 h2.inp`
`grep AUXIS h2.inp`
`grep - v AUXIS h2.inp`
`sort h2.inp`

writes the result of the command `cat` to the newly generated file `output` instead of to the terminal window. The command

```
cat h2.inp >> output
```

adds the result of the command `cat` to the already existing file `output`. Input files can be assigned with the < sign, but in many cases the < sign can be omitted and the input filename is written directly after the command and a space character.

It is possible to combine commands to new constructions called a "pipe." In a pipe construction the output of a command is used as input for the succeeding command. The command

```
grep AUXIS h2.inp| sort
```

extracts all lines containing the word `AUXIS` from the file `h2.inp` and sorts them alphabetically.

As long as the computer is working on a command, the shell does not show a prompt and new commands cannot be entered. If the sign `&` is added to the end of a command line, the processes are send to the background and the prompt returns immediately and is ready to accept new commands – this is very useful if programs that require more than just a few seconds of computing time are started.

```
myprog < input > output &
```

For file manipulations, a text editor is needed. The native editor on `Linux` systems is the `vi` editor. Unfortunately, the `vi` editor is not particularly friendly to new users, as it takes some time to get used to it. An easy-to-use alternative to the `vi` editor is the `nedit` editor, which is also available on the `Knoppix` Live; usage of the `nedit` editor is intuitively clear.

A.3
Character Tables for Chemically Important Point Groups

`Character Tables for Chemically Important Point Groups` is an Internet application for the analysis of molecular symmetry properties. The URL on the Internet is:
http://symmetry.jacobs-university.de

The installation on the `Knoppix` Live CD starts its own web server and does not need Internet connection. The application is accessible through the file viewer `Konqueror`. The application can be used for character tables, the reduction of reducible representations, force field analysis, and the determination of point groups.

On the application's starting page, a table with all chemically important point groups (currently 58 entries) is shown. Upon a mouse click on one of the point groups, the group's web page with its character table, additional group information, and symmetry properties for multipoles is generated. On each group's web page a dialog box for reducible representations can be found. With the button **Submit** the reducible representation is reduced to irreducible representations. The option Γ_{3N} performs a force field analysis up to quartic terms and determines the number of independent internal coordinates.

The option `Determine point group from molecular structure` on the application's starting page determines the molecular point group from either Cartesian or Z-matrix coordinates. The Γ_{3N} representation for the determined point group and the subgroups is generated automatically. Cubic point groups are currently not supported.

A.4
Computational Chemistry Software Delivered With This Book

A.4.1
`Molden`

`Molden` is a molecular editor and pre- and post-processing program for the visualization of molecular and electronic structure. `Molden` was developed at the Center for Molecular and Biomolecular Informatics (CMBI) at Radboud University, Nijmegen, The Netherlands, by Gijs Schaftenaar. The programs URL is:
http://www.cmbi.ru.nl/molden/molden.html

`Molden` can be used to visualize molecular structures, orbitals, electron densities, reaction paths, and vibrational modes.

The program can be started from a shell with the command `molden` or `molden input.mol`. The program launches two windows, the display window and the `Molden` control panel.

The Z-matrix editor can be started from the `Molden` control panel by clicking on the **ZMAT Editor** button. Entries to the Z-matrix are added through the button **Add Line**. The atom type is chosen from the periodic table window and the reference atoms are chosen by clicking on atoms in the display window. The program allows the modification of Z-matrix entries (element types, bond lengths, bond angles, or dihedral angles), the substitution of atoms by fragments, the measurement of bond lengths, bond angles, and dihedral angles, and many other features. It is recommended to spend some time with the Z-matrix editor to become familiar with all its features.

The quantum chemical program `deMon` can generate output in `Molden` format (file extension `.mol`). All relevant data for the visualization of calculated properties like orbitals, electron densities, and vibrational modes are written to that file. Orbital and density visualization is accessible through the **Dens. Mode** button from the `Molden` control panel. For vibrations click on the **Norm. Mode** button and select a frequency. The "Skull" icon terminates the program.

A.4.2
`deMon`

`deMon` is a quantum chemical program for the calculation of atomic and molecular properties. (The name originates from the place its first version was developed, Montréal, and stands for *densité de Montréal*, French for "Montreal's density.") `deMon` is based on density-functional theory. The program needs an input file with the extension `.inp`, and two example input files can be found on the `Knoppix` Live CD in the home account `examples` subdirectory. The program is started with the command

> `deMon file`

where `file` is the input file for the calculation. In the command line, the file extension `.inp` can be omitted and is added automatically. The program uses keywords to specify tasks. For the purpose of this book, only a few of the available tasks are used. The specification of the molecule's (starting) geometry in either Cartesian coordinates or Z-matrix is mandatory for all input files. If the output is visualized with the program `Molden`, the keyword `VISUALISATION MOLDEN` has to be used. For the calculation of thermodynamic properties with the `THERMO` scripts, a frequency analysis during the `deMon` run is necessary. A more complete `deMon` documentation can be found on the `Knoppix` Live CD and is accessible through the file viewer `Konqueror`.

The calculated results are written to an output file with the extension `.out`. Data to be interfaced with the program `Molden` are written to the file with the extension `.mol`.

Table A.7 contains the `deMon` keywords that are used for this book. Keywords can be placed in any order. If the program does not run, it usually cannot decode the input file. An error message is given in the output file. Lines starting with a hash (#) are ignored by the input parser.

A.4.3
CaGe

`CaGe` is a generator of mathematical graphs, and in the context of this book it is used for the generation of topological matrices for fullerenes. `CaGe` was developed at the Mathematics Department at Bielefeld University, Germany. The program's URL is:

http://www.math.uni-bielefeld.de/~CaGe

The program can be started from a shell with the command `cage.sh`. Only the option `general fullerenes` is used for the purpose of this book. In the next window the options `min=max` (default) and `Isolated Pentagons` (`IPR`) have to be selected, and the position of the slider determines the number of carbon atoms. If 60 atoms are selected, the famous buckminsterfullerene C_{60} is generated. The following window controls the output options, and it is recommended to select `3D representation` with viewer `Jmol Applet` and text viewer. Upon a mouse click on the **Start** button, the graph generation process is launched. Depending on the selected output options, several windows are opened after completion of the generation

Table A.7 deMon keywords used in this book.

Keyword	Description
GEOMETRY CARTESIAN	The geometry in Cartesian coordinates starts in the next line (in Å)
GEOMETRY Z-MATRIX	The geometry in *Z*-matrix coordinates starts in the next line (in Å and deg)
TITLE	Optional: Title for the calculation
VISUALISATION MOLDEN	The interface file .mol for visualization with the program Molden is generated
POPULATION MULLIKEN	Calculate charge distribution (Mulliken population)
DIPOLE	Calculate dipole moment
PRINT MOE	Print MO energies and occupation numbers
PRINT MOS	Print MO energies, occupation numbers and MO coefficients
CHARGE X	The total charge of the system is *X* (integer)
MULTIPLICITY X	The multiplicity $(2S + 1)$ of the system is *X* (integer)
OPTIMISATION	Perform a geometry optimization
FREQUENCY	Perform a frequency analysis (IR only)
FREQUENCY RAMAN	Perform a frequency analysis (IR and Raman)
THERMO	Calculate thermodynamic properties (requires also keyword FREQUENCY)
DYNAMICS MAX=X STEP=Y INT=Z	Run a molecular dynamics trajectory with *X* steps, a step size of *Y* (in fs) and give output in intervals of *Z* (in steps)
BATH NOSE T=X	Use a heat bath with the Nosé–Hoover chain thermostat. Request a temperature of *X* Kelvin
VELOCITIES ZERO	Start with all velocities set to zero
VELOCITIES RANDOM T=X LP=0	Start with random velocities that correspond to a temperature of *X* Kelvin. Request, however, that momentum and angular momentum of the initial structure are zero

Keywords specifying technical and numerical issues for the numerical solution of the electronic Schrödinger equation

BASIS (DZVP-GGA)	Use basis set DZVP-GGA
BASIS (SVP)	Use basis set SVP (single valence polarization, lower precision but higher performance than DZVP-GGA)
AUXIS (GEN-A2)	Automatically create auxiliary functions (numerical issue)
AUXIS (GEN-A2*)	Automatically create auxiliary functions with higher precision. Time consuming, but necessary for calculation of electron affinities of atoms
VXCTYPE AUXIS PBE	Use the auxiliary functions to compute the exchange correlation potential. Use the PBE functional
VXCTYPE AUXIS VWN	As above, but use local-density approximation instead. Less accurate, but about three times faster

process. In the **Result Selection** window, the button **adj** can be used to save the topological information. The Hückel program reads only the writegraph format, which is not the default setting for format. For a Hückel calculation, the number of p orbitals and p electrons has to be added manually (using an editor) to the writegraph output file in the first line. Information about the graph generation process is obtainable by a mouse click on the **log file** button.

If more than one isomer exists for a given number of atoms, the **flow** button can be used to complete the generation process for all isomers. The left and right arrows allow switching between isomers. The topological matrix can be saved again for each isomer via the **adj** button.

A.4.4
`hueckel`

We wrote a little Hückel theory program (`hueckel`) that can be used to perform HMO-type calculations. The program can be started from a shell with the command

```
hueckel < file
```

where `file` is the input file specifying details of the calculation. An example input file for the benzene molecule is available on the `Knoppix` Live CD in the home account `examples` subdirectory. The first line of the input file is used to specify the number of p orbitals contributing to the π system and the number of electrons. The following lines determine the connectivity (corresponding to the topological matrix) of the molecule. Each line stands for one atom. The first column specifies the atom number, and columns two to four specify the directly connected (bonded) atoms. If an atom has less than three neighbors, the lines have to be filled with zeros.

The `hueckel` program can read the `writegraph` format files from the `CaGe` program. If the `hueckel` input file is generated with the help of the `CaGe` program, the first line specifying the number of p orbitals and electrons has to be added manually.

A.4.5
THERMO Scripts

The `THERMO` scripts are a bundle of `bash` scripts for the calculation of thermodynamic properties. They were written by Robert Barthel at TU Dresden. The output file of a `deMon` calculation with frequency analysis and `THERMO` keyword is used as input for the computation of thermodynamic properties. The master script `THERMO.sh` calls the other scripts and the calculation is started with the command:

```
THERMO.sh file
```

where `file` is the output file of the `deMon` calculation.

A.4.6
Xmgrace

Xmgrace is a very popular program for data analysis from the open source community is started from a shell with the command

    xmgrace file

where file is the data file. It is recommended to use the extension .dat for data files, although the choice of an extension is completely free. The data file has one line per data point and two entries (xy pairs) per line for 2D representations. In the context of this book, Xmgrace is used to fit data of the potential surface to a Morse potential. This requires a nonlinear curve fitting. Create a new file (for example potential.dat) with the data for the potential surface, where x is the bond length and y is the total energy. After starting the program and loading the data, the option Data → Transformations → Non-linear Curve Fitting starts a new window. Select the data set from Source/Set and set the correct number of adjustable parameters. For the Morse potential, $y = (1 + e^{-\alpha x})^2$, the number of parameters is 1 and the correct entry for the Formula box is y= (1 + exp (-A0*x)) ^2. A mouse click on the **Apply** button starts the iteration process. It is recommended to use an appropriate starting value for A0 and to verify the result.

A.4.7
GNUplot

GNUplot is also a very popular open source program for data analysis. It is started from a shell with the command

    gnuplot

GNUplot uses a command line interface. For example, the command

    plot potential.dat with lines

reads data from the file potential.dat and connects the data points with lines. The complete GNUplot manual is available on the Knoppix Live CD and is accessible through the file viewer Konqueror.

Index

a
adiabatic IP/EA   144–146
adjacency matrix   72
affinity, electron, *see* electron affinity
algorithms
– Newton–Raphson   197
– velocity Verlet   172–173, 192
– Verlet   170–171
all-metal clusters   198, 202–205
– planar   147
ammonia molecule   21
– irreducible representation   138–139
amplitudes, vibrations   32
Andersen thermostat   187
angle
– dihedral   13–15
– torsion, *see* dihedral
angular frequency   29
angular momentum   108, 172–173
anharmonic oscillator   181
annealing, simulated, *see* simulated
    annealing
annulenes   67
antisymmetric stretching   114,
    123–124
– water molecule   131–132
approximations
– Born–Oppenheimer   69, 117, 169,
    199
– harmonic   25–27, 114–115
– ideal-gas   154–156
aromaticity, Hückel's rule   78
asymptotics, exponential   50
atmosphere, infrared spectrum   115
atomic energy levels   53
atomic masses, vibration
    amplitudes   32
atomic orbitals   47–55

– LCAO   7, 48, 69–70, 130
– orthonormalization   51
atoms
– electron affinity   57–66
– Schrödinger equation   47–49
– Slater rules   59–61
averages, thermodynamic   188
axes, rotational   15

b
bending   114, 123–124
– water molecule   131–132
Berendsen thermostat   186–187, 190
BFGS method   96
binding energies of electrons   59–61,
    74–75
bipyramidal structure   205
biradicals, stable   110
'black-box' strategy   1
bond length   13–14
– 1,4-difluorobenzene   136
– methane   161
– molecular dynamics   190
– reduced mass   28
– steepest descent method   98–101
bond order   83–84
bonding, fulvene molecule   89
Born–Oppenheimer
    approximation   69, 117, 169, 199
– simulated annealing   200
Born–Oppenheimer molecular
    dynamics (BOMD)   168
butadiene molecule   85–88

c
canonical ensemble   182, 186
Car–Parrinello molecular dynamics
    (CPMD)   168

Computational Chemistry Workbook: Learning Through Examples
Thomas Heine, Jan-Ole Joswig, and Achim Gelessus
Copyright © 2009 WILEY-VCH Verlag GmbH & Co. KGaA, Weinheim
ISBN: 978-3-527-32442-2

carbon atoms
– conjugated π systems   67–81
– sp² hybridized   85
carbon dioxide molecule   20–21
– normal modes   114
– thermodynamic properties   164
– vibrational modes   120
carbon monoxide molecule
– Hessian-based Newton–Raphson
  optimizer   100–102
– Morse potential   30
– point groups   20–21
– steepest descent method   98–100
– total energy   177
– vibrational frequency   43
Cartesian coordinates   11–12
– single-vector representation
  127–128
center of inversion   15
character tables, vibrational
  spectroscopy   127–141
charge order   84–85
chemistry
– computational
– thermochemistry   151–165
classical mechanics   35–37
classical molecular dynamics
  (CMD)   168, 174–178
classical vibrations, harmonic
  approximation   114–115
closed systems   173
clusters   147, 198, 202–205
computational chemistry   167–169
– basics   1–9
– simulated annealing   197–207
conjugated π systems   67–81, 83
conservation of physical
  quantities   172–173
coordinates
– Cartesian and internal   11–12
– mass-weighted   121
– single-vector representation
  127–128
Coulomb integral   71
critical point   155
cubic groups   17–18
cyanide, hydrogen   97, 124
cyclopropene, 1-methylene-   72
cyclopropenyl   75–78

d
degeneracy   131–132, 162
– orbitals   77–78
degrees of freedom   114, 136

– external   185
– molecular partition function   156
density, probability   38, 42
density-functional calculations   64
detachment energy, vertical   146
diamagnetic molecules   107
diatomic molecules   5
– forces   93–94
– geometry optimization   91–105
– Morse potential parameters   31
– vibrations   25–45
1,2-dichloroethene molecule,
  isomers   140
dichloromethane molecule, irreducible
  representation   138–139
differential equation, partial   39
1,4-difluorobenzene molecule
  136–137
difluoroethene molecule   92
– infrared and Raman spectra   125
– isomers   164, 200
dihedral angle   13–15
direct product, wavefunctions   133
discrete solutions   39
dissociation energy   26, 36
distortion, Jahn–Teller   109
double bond   84
dynamics, molecular, *see* molecular
  dynamics

e
EA, *see* electron affinities
eigenenergy   40
eigenfunctions   41
eigenvalues   76–77
– butadiene molecule   85
– translational and rotational
  motions   129
electron affinity   57–66
– adiabatic   144–146
– calculation   61–62
– influence of geometry   144–146
– molecules   143–149
– second period elements   65
– vertical   144–146
electron binding energy, π
  electrons   74–75
electron configuration   53
electron spin   107–112
electronic contribution to molecular
  partition function   160
energy
– atomic levels   53
– binding   59–61

– conservation   172–173
– dissociation   26, 36
– free   151–153
– internal   152–153, 163
– molecular orbitals   86
– orbital   48
– π electron binding   74–75
– potential energy surface, *see*
    potential energy surface
– total   172, 177
– units   164
– vertical detachment   146
– zero-point   36, 43, 152, 163
energy operator, kinetic   38
ensembles   182–183
– canonical   186
enthalpy   151–153
– reaction   164
entropy   153
– methane   163
equilibrium distance   189
equilibrium position   26, 30
ergodic theorem   183–185
ethane molecule
– irreducible representation
    138–139
– simulated annealing   205
ethene molecule, bond order   84
excitation operator   133
exclusion principle, Pauli   107
exponential asymptotics   50
extended Hückel theory (EHT)   67
external degrees of freedom   185
external heat baths   186–188
extrapolation, position vector   95

f
fluorine molecule   194
force constant   27
force-constant matrix   117, 121
forces, diatomic molecules   93–94
formamide molecule   19–20
– Z-matrix   20
formic acid   22
Franck–Condon principle   144
free energy   151–153
free enthalpy   151–154
frequencies
– analysis   139
– vibrational   28–30, 113
frontier orbital   86
fullerenes   67, 78
– isomers   98
fulvene molecule, bonding   89

g
gap, HOMO–LUMO   73
gas phase, relative abundance of
    isomers   160–161
Gaussian function   41
geometry
– influence on IP and EA   144–146
geometry optimization   3, 33
– all-metal clusters   202–205
– diatomic molecule   91–105
– 1,4-difluorobenzene   136
– global   97–98
– local   97–98
– methane   161
– Morse potential   97
Gibbs' equations   155
global geometry optimization
    97–98
global maximum/minimum   92–93,
    203–205
gradients, calculation   173
grand canonical ensemble   182

h
Hamilton integral   69
Hamiltonian operator   37
– off-site matrix elements   71
harmonic approximation   25–27
– classical vibrations   114–115
harmonic oscillator   115–117
– probability density   42
– quantum mechanical   35
– wavefunctions   40
harmonic potential   27
harmonics, spherical   59
Hartree unit   60, 62
heat bath   174
– external   186–188
heat capacity   156
– methane   163
Hermite polynomials   41
Hessian-based Newton–Raphson
    optimizer   7, 95–96
– carbon monoxide molecule
    100–102
Hessian matrix   94, 96, 119–121
– character tables   127–129
highest occupied molecular orbital
    (HOMO)   58, 61
HOMO–LUMO gap   73, 86–87
– stability of polyenes   88
Hooke's law   27, 115
horizontal mirror plane   17
Hückel postulates   70–71

Hückel theory 7, 67–90
– bond order 83–84
– charge order 84–85
– extended 67
– molecular orbitals 83–90
– Schrödinger equation 68
Hückel's rule for aromaticity 78
Hund's rule 77–78, 109–110
sp^2 hybridized carbon atoms 85
hypersurface, potential energy 92

i

ideal-gas approximation 154–156
ideal-gas equation 154
in-plane bending 114, 123–124
inactive vibrations 120
inertia, moment of 159
infrared spectroscopy
– atmosphere 115
– difluoroethene 125
– intensities 119–120, 133
– selection rules 133–135
internal coordinates 11–12
internal energy 152–153
– methane 163
inversion, center of 15
iodide, hydrogen 103
iodine molecules 103
ionization potential (IP) 57–66
– adiabatic 144–146
– calculation 61–62
– influence of geometry 144–146
– molecules 143–149
– second period elements 64
– vertical 144–146
ions, Slater rules 59–61
irreducible representation 131–134,
138–139
isolated pentagon rule 78
isomers 92
– 1,2-dichloroethene 140
– difluoroethene 164, 200
– fullerenes 98
– relative abundance 160–161
– simulated annealing 197
iteration 99–100, 172

j

Jahn–Teller distortion 109
Jahn–Teller effect 108–110

k

kinetic energy 172
– macroscopic systems 183

– operator 38
Koopmans' theorem 61, 147

l

laws and equations
– ergodic theorem 183–185
– Franck–Condon principle 144
– Gibbs' equations 155
– Hooke's law 27, 115
– Hückel's rule for aromaticity 78
– Hund's rule 77–78, 109–110
– ideal-gas equation 154
– isolated pentagon rule 78
– Koopmans' theorem 61, 147
– Maxwell–Boltzmann
distribution 185
– Morse potential 5, 25
– Newton's law 27, 116
– Newton's second law 172
– Pauli exclusion principle 107
– Schrödinger equation, *see*
Schrödinger equation
– Stirling formula 155
Lennard-Jones potential 198
linear combination of atomic orbitals
(LCAO) 7, 48, 130
– Hückel theory 69–70
linear groups 17–18
linear molecules, vibrational
modes 120–124
local geometry optimization 97–98
local maximum/minimum 92–93,
198
lowest occupied molecular orbital
(LUMO) 61
– *see* also HOMO–LUMO gap

m

macroscopic systems
– kinetic energy 183
– thermodynamics 185–186
many-body quantum mechanics 2
mass
– reciprocal 174
– reduced 27–28
mass spectrometry 143
mass-weighted coordinates 121, 128
matrix
– adjacency 72
– force-constant 117, 121
– Hessian, *see* Hessian matrix
– topology 71–72
– Z-matrix 12–13, 19–20, 191,
202–203

maximum, global/local 92–93
Maxwell–Boltzmann
    distribution 185
MD, *see* molecular dynamics
mechanics, classical/quantum 35–37
mesomeric structures 75
metal clusters 198, 202–204
methane molecule
– Cartesian and internal
    coordinates 14–15
– geometry optimization 161
– thermodynamic properties
    161–164
– Z-matrix 15
1-methylene-cyclopropene
    molecule 72
microcanonical ensemble 182, 191
minimum, global/local 92–93, 198,
    203–205
mirror planes 15–17
modes
– normal 114, 129–132
– vibrational 117–124
molecular coordinates 11–23
molecular dynamics (MD) 9
– algorithms 3
– basic concepts 167–179
– basic thermodynamics 181–196
– classical 168, 174–177
– energetics 178
– hydrogen chloride 178
– simulated annealing 197–207
molecular orbitals 51
– 3D 53
– energy levels 86
– highest occupied, *see* highest
    occupied molecular orbital
– Hückel theory 7, 67–90
– lowest occupied, *see* lowest
    occupied molecular orbital
– π electrons 74
– theory 68–70
– theory, *see also* Hückel theory
molecular partition function
    156–160
– degrees of freedom 156
molecular point groups 20–21
– carbon monoxide/dioxide 20–21
molecular units 154
molecules
– ammonia 21, 138–139
– butadiene 85–88
– carbon dioxide, *see* carbon dioxide
    molecule

– carbon monoxide, *see* carbon
    monoxide molecule
– diamagnetic 107
– 1,2-dichloroethene 140
– dichloromethane 138–139
– 1,4-difluorobenzene 136–137
– difluoroethene 92, 125, 164, 200
– electron affinity 143–149
– ethane 138–139, 205
– ethene 84
– fluorine 194
– formamide 19–20
– fulvene 89
– hydrogen 103
– hydrogen chloride 178
– hydrogen cyanide 97, 124
– hydrogen iodide 103
– iodine 103
– ionization potential 143–149
– linear 120–124
– methane 14–15, 161–164
– 1-methylene-cyclopropene 72
– nitrogen monoxide 110
– ozone 147
– paramagnetic 107
– planar 22
– water, *see* water molecule
moment of inertia 159
momentum
– angular 108
– conservation 172–173
Morse potential 5, 25
– carbon monoxide 30
– diatomic molecules 31
– geometry optimization 97
– Schrödinger equation 36
– thermochemistry 174–176
– thermodynamic averages 188
multi-dimensional problems 117
multiplicity 108

n
Newman projection 14
Newton–Raphson algorithm 197
Newton–Raphson optimizer 7,
    95–96, 100–102
Newton's law 27, 116
Newton's second law 171
nitrogen monoxide molecule 110
nodes 52, 54
non-axial point groups 17–18
normal modes 114
– carbon dioxide molecule 114
– character tables 129–130

– symmetry   130–132
– visualization   140
normalization   38, 48, 50
Nosé–Hoover thermostat   187–188
NP-hard problems   198
nuclear charge   59
nuclear fission   190
nucleus–nucleus repulsion   68
NVE ensemble   182
NVT ensemble   182, 186

o
off-site Hamilton matrix elements
   71
open systems   173
operations, symmetry   16–17
optimization, geometry, *see* geometry
   optimization
optimizer, Newton–Raphson   7,
   95–96, 100–102
orbital energy   48
orbital overlap   71
orbital shape   51
orbitals
– atomic, *see* atomic orbitals
– degenerate   77–78
– frontier   86
– molecular, *see* molecular orbitals
– valence   53
orthogonality   48
orthonormality   43, 51
oscillator, anharmonic   181
out-of-plane bending   114,
   123–124
overlap, orbital   71
overlap integral   69
1,2,5-oxadiazole   22
ozone molecule   147
π electron binding energy   74–75
π systems, conjugated   67–81
– conjugated   83

p
paramagnetic molecules   107
Parrinello, *see* Car–Parrinello
   molecular dynamics
partial differential equation   39
particle spin   49
partition function   152–154
– molecular   156–160
Pauli exclusion principle   107
PES, *see* potential energy surface
phenyl rings   110
planar all-metal clusters   147

planar molecules   22
Planck constant   36
point, critical   155
point groups   17–19
– carbon monoxide/dioxide   20–21
– character tables   134
– molecular   20–21
– non-axial   17–18
– symmetry numbers   159
polyenes, stability   88
polynomials, Hermite   41
position vector, extrapolation   95
potential
– harmonic   27
– ionization, *see* ionization potentials
– Morse, *see* Morse potential
– pair   198
– pseudo-potential   63
potential energy   172
– thermochemistry   151
potential energy surface (PES)   5, 25,
   91–93
– simulated annealing   200
– thermochemistry   174–176
principal quantum number   49, 60
probability density   38
– harmonic oscillator   42
– π electrons   74
product, direct   133
projection, Newman   13–14
propagation of atoms   169
pseudo-potential   63

q
quantization   40
quantum chemical methods   2
quantum mechanics   35–37
– harmonic oscillator   35
quantum molecular dynamics
   (QMD)   168
quantum number   48

r
radical, triphenylmethyl   108
Raman spectroscopy   115
– difluoroethene   125
– intensities   119
– selection rules   133–135
reaction enthalpy   164
reciprocal mass   174
reduced mass   27–28
reference atom   13–14
relative abundance of isomers
   160–161

representation, irreducible   131–134
repulsion, nucleus–nucleus   68
rings, phenyl   110
rotational axes   15
rotational contribution to molecular
   partition function   158–159
rotational motions, eigenvalues   129
σ/π separation   70

s
saddle point   93
scaling thermostat   186
SCF, *see* self-consistent field technique
Schrödinger equation   5–6, 33, 35–45
– atom   47–49
– Hückel theory   68
– simulated annealing   201
– solutions   39–42
– stationary   37–39
– thermochemistry   151
screening constant   59–61
second period elements, IP and
   EA   64
selection rules   133–135
self-consistent field (SCF)
   technique   111
shielding constant   60
simple scaling thermostat   186
simulated annealing   197–207
– potential energy surface   199–200
– stationary points   202–203, 205
simulations   1, 167–169
single-vector representation,
   Cartesian coordinates   127–128
Slater rules   59–61
spectrometry, mass   143
spectroscopy
– infrared/Raman   115, 119
– simulation   1
– vibrational   113–141
spherical harmonics   59
spin
– angular momentum   108
– electron   107–112
– particles   49
spreadsheets   174–177
stability
– biradicals   110
– conjugated carbon π systems
   67–81
– polyenes   88
starting point, iterations   103
startup conditions   169
stationary points   202–203, 205

stationary Schrödinger equation
   37–39
statistical thermodynamics   152
– ergodic theorem   184
steepest descent method   94–95
– bond length   98–101
– carbon monoxide molecule   98–100
step size   100–102
Stirling formula   155
stretching   114, 123–124, 131–132
– antisymmetric, *see* antisymmetric
   stretching
structural isomers   92
surface, potential energy, *see* potential
   energy surface
symmetric stretching   114, 123–124
– water molecule   131–132
symmetry
– basics   15–16
– elements and operations   16–17
– molecular coordinates   11–23
– normal modes   130–132
– water molecule   130–132
symmetry elements, water
   molecule   16
– water molecule   132
symmetry numbers   159
system size   148

t
tables, character   127–141
Taylor series   26
temperature   183–185
tensor of moments of inertia   159
thermochemistry   151–165
– ideal-gas approximation   154–156
thermodynamic functions   152–154
thermodynamic properties
– carbon dioxide molecule   164
– methane   161–164
thermodynamics
– averages   188
– ensembles   182–183
– macroscopic systems   185–186
– molecular dynamics (MD)   181–196
– statistical   152
thermostats   174, 186–188
– Berendsen   186–187, 190
topological structures   201
topology matrix   71–72
torsion angle   14
total (angular) momentum   172–173
total energy   172
– carbon monoxide   177

trajectory   169
transformed model system   29
transition states   202
translational motions   157–158
– eigenvalues   129
triphenylmethyl radical   108

u
units
– energy   164
– molecular   154

v
valence orbitals   53
velocity Verlet algorithm   171–172, 192
Verlet algorithm   170–171
– velocity   192
vertical detachment energy   146
vertical IP/EA   144–146
vertical mirror plane   17
vibrational contribution to molecular partition function   159–160
vibrational frequencies   28–30, 113
– carbon monoxide   43
– 1,4-difluorobenzene   138
vibrational levels   36
vibrational modes   117–124
– carbon dioxide molecule   120
– linear molecules   120–124
– water molecule   131–132

vibrational spectroscopy   113–126
– character tables   127–141
– intensities   119–120
vibrations
– diatomic molecules   25–45
– harmonic approximation   114–115
– wavenumber   32
visualization, normal modes   140

w
water molecule
– Cartesian and internal coordinates   11
– symmetry   130
– symmetry elements   16
– vibrational modes   131–132
– Z-matrix   12–13
wavefunctions   38
– direct product   133
– harmonic oscillator   40
– Slater rules   59–61
wavenumber, vibrations   32

z
Z-matrix   12–13, 19–20
– formamide molecule   20
– methane molecule   15
– molecular dynamics   191
– simulated annealing   202–203
– water molecule   12–13
zero-point energy   36, 43, 152, 163